The Gasoline Age-Why?

Herbert J. Redman

Bloomington, IN   authorHOUSE™   Milton Keynes, UK

AuthorHouse™
1663 Liberty Drive, Suite 200
Bloomington, IN 47403
www.authorhouse.com
Phone: 1-800-839-8640

AuthorHouse™ UK Ltd.
500 Avebury Boulevard
Central Milton Keynes, MK9 2BE
www.authorhouse.co.uk
Phone: 08001974150

First published by AuthorHouse 7/26/2006

ISBN: 1-4259-3635-0 (sc)
ISBN: 1-4259-3636-9 (dj)

Library of Congress Control Number: 2006904446

Printed in the United States of America
Bloomington, Indiana

This book is printed on acid-free paper.

ACKNOWLEDGEMENTS

In a work of this magnitude, there are always contributors who put forth their time, and, often their money to help further the cause of historical studies. First, I would like to acknowledge the enormous debt I owe to two of the giants in the field of automotive history, Beverly Rae Kimes and Keith Marvin. Both were very instrumental in helping me to retrieve valuable hard-to-find information that greatly facilitated in the preparation of this work. It is my sincere hope that any new research facts unearthed during the long searching out process can only enhance the understanding of this oft overlooked topic from these trailblazers.

There are of course, many others who helped out on this quest. They are listed below (not in any particular order, I might add): Beth Michael and the reference staff of the Defiance Public Library; John R. Link of Indiana, who is related to one of our pioneers and provided much valuable information; Earl Mather of Garfield, Ohio; William Lewton of Lisbon; the staff of the Springfield Public Library; Stuart McDougal from the Philadelphia Free Library, who never failed to answer the call; Gregg D. Merksamer of New York State, who encouraged researching the production figures among the alternative cars in the early years as well as the Madison Square Garden showcases in those pioneer days; the staff of the Ross County Public Library in Chilicothe, who answered numerous questions; the Ross County Historical Society and Patricia Medert; Randall Grabb of Chillicothe, who went out of his way to help uncover the elusive Gramm Steamer; the Stark County Historical Society, located in Canton; the McKinley Museum, also in Canton; Stark County District Library, especially Debra Pfender and Mark Guyer; Dayton & Montgomery County Public Library; Salem Public Library of Salem; Paul Stanley of the Damascus Historical Society; finally, but certainly not least, Laura Corwin and the staff of the National Automotive History Collection at the Detroit Public Library. The fine people who work there have never failed to fulfill any of my strange requests for information, including old periodicals from the time periods we are concerned with in this work. Quite a few of those periodicals I might add.

To any I may have inadvertenly left out, I apologize sincerely. On a more personal note, I wish to thank my wife, Vicki, and our two children, Hannah and Joshua. Their support in this much larger project than I ever thought it would be was essential.

Table of Contents

First Marque   The Aultman Steam of

Canton, Ohio. (1901-1902) 10

Second Marque. The Gramm Steamer of

Chillicothe, Ohio. (1902-1903) 25

Third Marque   The Krotz Electric of

Springfield, Ohio.(1902-1905) 35

The Krotz Gas-Electric of

Defiance, Ohio.(1908-1911) 35

Fourth Marque   The MacDonald Steamer of

Garfield, Ohio.   (1923-1924) 50

Fifth Marque The Thresher Electric of

Dayton, Ohio.(1900) 64

The Final Analysis   The Failure of the Alternative   Cars?I.

Automobile History Often Shrouded In Uncertainty. 92

CHAPTER ONE

There were *three, then there* was *one.*

Few of mankind's many and varied inventions has had such a long-ranging, all-encompassing impact on modern history as has the automobile. This certainly applies from the time the earliest automobiles began to appear in noticeable numbers on America's roads from the mid-1890's on. Additionally, few have gone through as many different permutations on the same basic theme as the car. Yet, barring the "perfection" of the long-awaited fuel cell, the basic automobile structure appears set, at this writing, with the usual minor deviations, into the foreseeable future.

This subject touches everyone. Granted, not every person who *could* drive does, but this modern conveyance has become so much a part of our everyday lives it would likely now be impossible to imagine society without it. Further, a large part of this country's workers make their living from the motorcar, either directly or indirectly. The automobile has reached the stage where it is taken for granted. As such, it is not odd to see the dearth of academic writing on this topic.

That aside, the cars of the end of the twentieth century were very much as they had been for decades, with the same basic type internal combustion engines. Yet, like the cars it powered, this engine type, too, had gone through several metamorphoses by that point; like twin-cams, superchargers, and various cylinder configurations ranging from in line threes and fours through V10s. But, still the same basic engine designs. Although electric cars, like the EV1 from General Motors, were available, they were meant for a limited market with very little chance for expanded production. The existence of this exception, if anything, by that point had proven the rule about the gasoline engine's dominance in the automotive world.

Yet, there was no denying the fact finite supplies of fossil fuels, coupled with their increasing costs—not to mention the public's continuing perception of the gasoline engine as a polluter—had spurred forward development of alternative powered cars. There was an increasing public call for alternative-powered, "environmentally friendly" cars. However, a viable, mass market "speciality" vehicle remained elusive. The turn of the century did find engineers working hard to tackle the problems like never before. Incidently, in this case, alternative meant any car not solely powered by internal combustion gasoline engines.

With these alternative cars poised for a possible comeback, the turn of the century also seemed like a good time to be investigating why they had failed to start with. To be sure, the contemporary progress of alternative cars was not of historical concern then. However, as the nation embarked upon the second century of the American automobile, it was certainly worthwhile to look over why these power sources had failed at the beginning of the first.

In this work, we will examine those early days when both steam and electric were distinct possibilities as power sources for the then fledgling car industry. When they had just as much of a chance as the gasoline engine. We will endeavor to explain why the alternative cars developed in the manner they did, and, more importantly, why they ultimately failed to satisfy the American consumer. We will do this, in part, by examining five of the early alternative cars found at or near the beginning of the twentieth century that were also made in Ohio. Why we are interested in the time frame has already been established. But, 'Why Ohio?' 'Why not Indiana or Kentucky or any other state where cars were made?'

Indiana, home of the annual Indy 500 auto race, has had a relatively generous amount of coverage for its rather widely scattered automobile industry. Kentucky never produced more than a few cars, all of them rather ordinary. Ohio, in sharp contrast to the numerous studies of the car industry in neighboring states like Indiana and Michigan, has seen its own industry largely ignored. This is rather surprising, especially since all three states played integral roles in the early American auto industry, and were easily the center of car production. For much of the twentieth century, in fact, Ohio was second among all states behind only Michigan both in terms of output and number of individual makes.

Moreover, Cleveland, destined to be the center of electric car production for a time, was also the first "Motor City," before Detroit so unceremoniously wrestled away that title. It is noteworthy a greater variety of different model electrics, as well as larger numbers, were produced in Cleveland-based factories than anywhere else in the country. But, an excellent, although concise, study of Cleveland cars exists already; to do another would be just duplicating a fine effort. Nevertheless, even well-known gasoline-powered marques like Peerless of Cleveland have yet to be the subject of an exhaustive survey, while for at least one case—that of the Aultman Steam car of Canton, Ohio—there has been incorrect information published in a well-known national old car publication. In the present work, we will correct this oversight with the Aultman, as well as look over four other obscure alternative cars made here in Ohio.

Admittedly, Ohio is a strictly artificial choice for our purposes, but we must set a limit to the study somewhere. In sum, despite its early prominence in the car manufacturing field, Ohio's industry has been largely ignored. The following, in part, is an effort to redress that imbalance. What else do we hope to accomplish from all of this?

There has never been, to the knowledge of this writer, more than a cursory examination of this whole issue of why the alternative cars ultimately fell before the relentless advance of the internal combustion engine. When this query was made at the National Automotive History Collection, housed in the Detroit Public Library, there was surprisingly little information available. Nor is there a shortage of first-rate, quality books and periodicals devoted to this important industry, as a brief examination of the bibliography will make clear to the reader. However, few make more than a passing nod in the direction of this whole issue, with the limited information provided therein about the failure of the alternative car types often simply repeated over and again in the consulted sources. With that in mind, we will use the second part of this study to briefly examine this whole issue. We are thus touching on two separate, but related, topics in this thesis. However, neither one has yet received the coverage it so individually deserves from historians.

There is another question here: 'Why not include gasoline cars from at least Ohio and many additional makes?' For a couple of reasons, but a main one is because a good study, we will note, must define clear parameters. Obviously, covering a history of the dozens of Ohio makes, to say nothing of the national output, would be impractical, time-consuming, and beyond the limits of a study like this. Additionally, hundreds of gas marques may have failed, but the gas car itself endures. Our modern darling, the internal combustion engine, has been the subject of much scrutiny and experimentation to bring it to the point it is currently at. In short, we know what *it* can do. We do not similarly understand the capability of the alternatives. Especially since the period they were in vogue for was so brief and so remote in time.

The makes covered here have been selected because the involved concerns produced genuine product, with the hopeful expectation of volume production and, ultimately, a successful business effort. There are, among the five, both steamers and electrics, even a gasoline-electric dual power car. In every case, production was a genuine plan, and there has been enough evidence to say real cars emerged in each case. This precaution was taken to preclude "pie in the sky" dreamers or impractical designs from making this short list. Nevertheless, one of the cars was built by a businessman who would spend time in jail for fraud in connection to a previous marque. Despite this, his later marque appears to have been a genuine, bona-fide effort.

Now, as we have observed, the internal combustion engine is certainly predominant in the car industry today. So predominant we often forget this has not always been the case. It is not commonly thought of today, but, "from 1900 to 1908, there were twice as many steam cars on American roads as gasoline autos."[1] Of course, part of the reason the gas car made it was because of inherent flaws in the steam and electric cars, which we will get to later on. One indisputable fact remains, though, steamers and electrics were already highly developed, and much more sought after, at the turn of the twentieth century than their gas counterparts. Production of American cars of all three types, from 1900, bear this out. There were 1681 steamers, 1575 electric cars, and just 936 gasoline cars made in that year.[2] In fact, this would mark the zenith year for alternative car production in terms of percentage. Moreover, in 1893, when the gas powered Duryea car was first being put to the test on the open road, "electric carriages were already 'for hire' at the Chicago World's Fair."[3] We can get a glimpse from the preceding just how big an impact these early alternative cars were making before the coming of the gas car spoiled the "honeymoon." Yes, the alternative cars did get a big head start, we can not deny. Yet, in 1904, only a short time after, 18,500 gas cars were made in America, but only 2993 steamers and some 649 electric cars.[4] From that point on, the gas car would command the lion's share of production.

This intriguing situation needs closer examination. Two important questions come to mind, immediately. 'How did the internal combustion engine so decisively win the battle for America's consumers?' Not only that, but, 'How did this happen in such a short time?' We will look for the answers in this query.

It is appropriate at this point to look over the principal points of this writer's argument. They are: 1.) It is often asserted the alternative cars were defeated by the arrival of the Self-Starter on gasoline cars. The Kettering starter arrived relatively late, for the 1912 model year. This writer contends the alternative cars had already been defeated in essence *years* before this. That the Self-Starter's introduction simply ensured the alternative cars would not make an immediate comeback; 2.) Although initially the steamers and electrics were technologically superior to the first crude gas "buggies," this writer contends that imbalance was quickly corrected in favor of the gasoline cars. In short, the internal combustion engine speedily overtook the others, outstripping them in both popularity and technology early on; 3.) Finally, it is often zealously, even religiously, postulated the alternative cars got a "raw deal"; that they were assigned to oblivion even when shown to be superior. Even that there was a conspiracy to put them out of business. This writer sets out to show this was not the case at all. He contends both alternative types (and any variation thereof) had rather quickly displayed critical shortcomings that showed their inadequacy as suitable power sources for the first cars. He is confident the following will provide the solid basis for this assertion.

First Marque
The Aultman Steam of Canton, Ohio. (1901-1902)

At the turn of the last century, 30,667 persons called Canton home.[5] If existing accounts are to be believed, and there is no reason to doubt their authenticity, this northern Ohio city became an automobile center very quickly. Even before the turn of the century, this was apparently the case.

The first car in town[6] was owned by a Jebulon W. Davis; it was a Winton gas car, haling from Cleveland.[7] This was about the winter of 1897-1898. It was about the same time, no doubt, when enterprising local firms/individuals started dreaming of building their own cars. One concern that did just that went by the name of Aultman.

Reminders of the Aultman name existed, at this writing, all over Canton, most notably by the Aultman Hospital. Once upon a time, though, the city was also the home of C. Aultman & Co., founded by Cornelius Aultman (1827-1884) in 1851, a manufacturer of agricultural implements. This was the firm that would eventually make the Aultman Steam.

This diminutive vehicle, the first of the little-known marques we will consider in this thesis, was not the only car "made" in Canton.

In fact, it is just one of seven marques that once called Canton home. To be sure, most of these were of the ordinary, bland type of early gas buggy not likely to warrant more than a passing glance.

Not so the Aultman Steam. That particular car is tinged with controversy. And it was most definitely an alternative car, "designed to be a competitor to the Stanley Steamer."[8] It happened that its manufacturer was in desperate need of a saleable product about then.

With the firm's principal business of mowers and threshers suffering from a growing field of new competitors, and the old executives who had so successfully guided it having all taken what can be euphemistically called 'permanent vacations,' Aultman & Co. was in real trouble by the 1890's. Entering the brand new field of automobiles was just one of the ventures the company was trying. All in the name of survival.

Now to the controversy. The monumental *Standard Catalog of American Cars 1805-1942*, at this writing in its third edition, is a fine work, containing much information not readily available elsewhere. However, this reference work had not cleared up a mystery unearthed early on in regard to the Aultman Steam.[9]

On page 30 of this reference, a Henry J. Altman is listed as living at 11 Pier Street in Cleveland. Now, on page 78, another Henry J. Aultman (note the spelling), "built his first experimental steam carriage in Cincinnati in 1898 and moved to Canton in 1901 to establish the Aultman Company for its manufacture."[10] Could there be a connection?

A check of the Cleveland City Directories for 1901-1902 turned up some additional evidence. There is one reference for a Henry J. Aultman and another for the above cited Henry J. Altman, both in Cleveland. Here is the rub. *Both* are listed at 11 Pier Street in Cleveland. In the 1902 directory, Henry J. Aultman's occupation was listed as a trimmer.[11] As for Altman, his occupation, in living at this same address, is listed as an "electr'n" (electrician).[12]

Could one of these two gentlemen—assuming there were two— be the creator of the Aultman Steam? A little genealogical research was in order here. In fact, this is an excellent example of the research problems inherent in the history of the car.[13]

Another problem: the Aultman Company in Canton was already a going concern by 1901. Founded by Cornelius Aultman, as we have already observed. Nor is this all. A look at contemporary Canton directories yielded some additional surprises. The Canton City Directory for 1902 does not list *any* Altmans. The 1901 directory has one.[14]

What about the alternative 'Aultman' spelling? There are a few listings.[15] But, significantly, no Henry J. Aultman is to be found.[16] The 1902 directory lists three Aultmans, besides the firm.[17] But, again, no Henry J. Aultman. In fact, the Aultman family was no longer in control of the firm by that point.[18]

As for any Aultmans living in Cincinnati about this time, a check of city directories for 1898, 1899, and 1900 did not turn up persons using *either* form of the spelling. This is odd since this gentleman was supposed to have resided in Cincinnati about this time. Now the question of importance becomes, 'Are these two gentlemen in fact one and the same?'

Further, *The American Car Since 1775* fails to mention Cincinnati as the home of any Altman/Aultman cars. There *may* have been a Henry J. Altman (or Aultman) who lived in Cincinnati around the turn of the century who built a car, and chose not to make that fact known. But, he has not been found in the record. Besides, that discrepancy would be difficult to explain in terms of the free advertisement such a listing would entail.

How about Cleveland? Richard Wager's *Golden Wheels*, a history of cars built in and around Cleveland, likewise does not mention either spelling at all.[19] In addition, *The American Car* could not clear up the issue. In the listing for Aultman, in the section entitled "5000

Marques," the compilers use Cleveland as the base for production in 1898, then Canton in 1901-1902.[20] Certainly more than one car emerged from all of this. As for the Altman built by Henry J., it was indisputably a one-off car. That much *is* known.

As for the controversy over whether there was more than one gentleman, the only conclusion can be, while there certainly was a Henry J. Altman alive and living in Cleveland at the turn of the century, he had **nothing** whatever to do with the Aultman firm in Canton.

To support this contention, a check of the 1905 Cleveland City Directory turned up a few more points. There is our Henry J. Altman listed, residing then at 859 Fairmount Street, with engineer noted as his occupation.[21] The point is, Mr. Altman was listed as *still* living in Cleveland at that point, not in Canton working with the Aultman Co.; which, by then, was closing up. As *Standard Catalog* states, "irrevocable receivership arrived in September 1905."[22]

So we can find one of the "two" men we look for, and can not even establish the other's existence with certainty. Of course, this is just one little facet of research, but a thought provoking one.

Beverly Rae Kimes, in the Notes on the Third Edition of *Standard Catalog* states, "what the preparation of this book has demonstrated is how severely we have underestimated the extent of automobile building in this country." This writer submits, in view of the foregoing, if there were subtle errors here in the Aultman case, there were bound to be many others in automotive history as it stood at this writing. Besides, there are yet more twists to this storied tale.

As for the steamer itself, the marque is briefly mentioned in Healds' *The Stark County Story.* Heald relates, "The Harry Stuts automobile shop of Springfield, Ohio, was bought up, and Harry Stuts himself brought to Canton."[23] Here, again, we have real inconsistencies. This information insists a man named Harry Stuts was living in Springfield and apparently doing car work of some kind around the turn of the century. But, a check of Springfield city directories for 1900 and 1901 do *not* list a Harry Stuts anywhere. Likewise, there is nothing to connect any such individual to an automobile shop thereabouts.

Could this individual, in fact, have been the same Harry Stutz for whom the Stutz cars were later named? Does not seem likely, since the whereabouts of Harry C(layton) Stutz about this same time are well documented. The latter would have been in Dayton, however, about the same time as Harry Stuts was in Springfield. A check of Dayton City Directories for 1900 and 1901—the two years are combined in one volume—do indeed show a Harry Stutz.[24] In later years, the information generally stayed the same, except Harry's occupation was by then listed as "machine shop."

In any case, the Harry Stutz we know of opened a small machine shop in Dayton, "putting in all of his time trying to perfect a gasoline motor for stationary and automobile work."[25] There are no known references to place *this* Harry Stutz in Springfield for any long-term endeavor, business or otherwise, at any point. As for the Harry Stuts who went to Canton, he "designed a gas car, which the late Ted Reefsnyder helped to build, patterned after a single cylinder English gas motor."[26] This may have been the "tiller steered gas buggy" mentioned in *Standard Catalog,* although the subsequent reference of a move to Indianapolis, "where it went into production," seems muddled.[27]

Now, once more, there are problems galore. When we looked for this Theodore Reefsnyder, he was not listed in Canton city directories from 1898 through 1905. Moreover, a check of the 80-plus makes of cars once manufactured in and immediately around Indianapolis shows only one Harry Stuts/Stutz involved anywhere: the aforementioned Harry Stutz of the Stutz cars. Harry Stutz did indeed move to Indianapolis, as we have already observed, and as per Heald's reference.[28]

Now we can examine what we know irrefutably. According to Canton directories from the early 1890's to 1903, a William Lynch was affiliated with the firm; his address was listed as 420 S. Market Street.[29] We need to get this correct. For, if the foregoing were all true, William Lynch was the "father" of the Aultman Steam, not any Henry J. Aultman.

When William Lynch stepped up to the helm, however, the old firm was in deep trouble. He dug in, and early on resolved to diversify, so that, "the factory would operate continuously throughout the year, the purpose of which was to increase profits and incidently provide year round employment."[30] According to Heald, "heading the Aultman Co.. . . [then] were William A. Lynch, prez.; Ira M. Miller, vice-prez; Jay M. Cogan, Secretary."[31] Coming into the picture with the new management were some new, and desperately needed, funds. Lynch and his associates had divested themselves of some interests in local streetcar concerns, as well as a substantial block of stock in The Canton, Light, Heat [,] and Power Company, of which Lynch had been vice-president.

All of these transactions left these businessmen with a cool $400,000, most of which was promptly invested in the Aultman Company. A point of possible confusion here needs clarified. A firm called The Aultman & Taylor Company of Mansfield, Ohio, although founded in part by Cornelius Aultman, bore no connection at any time with the company in Canton. It *never* had any connection to the Aultman Company's automotive endeavors.

Now, to the steamer. There is surprisingly little available in the local area on the Aultman Steam. Visits to the Stark County Public Library produced little beyond the city directories referred to earlier.[32] An effort to uncover E. T. Heald's source materials was undertaken, but the original interview notes that the author had used years before in the chapter's preparation were not found.[33]

Nonetheless, we can piece the steamer's story together utilizing the various fragments unearthed in this research. Directory listings about the products of C. Aultman & Co. for 1898 detail them as, "mfrs steam threshing machinery, American threshers, Star road engines and stackers, clover hullers, water tanks, grain weighers and baggers, horse powers, road making machinery, oil engines, elevating and conveying machinery, etc., office and works 920 S. Market."[34] The 1899 directory repeats this listing, almost verbatim.[35] The 1901 listing reflects Lynch's commitment to diversification. Along with the host of other products, "automobiles" are there listed for the first time.[36] But, in the business index in the back of the same directory, C. Aultman & Co. is not listed in the ranks of automobile manufacturers. There, the only listing is of a T. S. Culp.[37]

It may well have been that Aultman commenced auto manufacture in 1901 too late to be included in the business directory section, for in the 1902 business section, Aultman is listed along with Culp as an auto maker.[38]

The details are sketchy.[39] But what emerges from the literature, after a strenuous search, seems fairly complete. *The Horseless Age* shows the above cited photograph and relates: "The new steam carriage which the Aultman Company, Canton, Ohio, are now prepared to deliver."[40] Given the lead time for periodicals, which at press time showed cars being prepared for delivery to customers, but not yet ready, could explain the 1901 business listing cited above. Significantly, the same source details the firm's experiments "with a steam truck having four-wheel drive for some time," but that, "such efforts have been put on hold while the car was undergoing trial." *Standard Catalog* relates, "Further experimentation ensued with a four-wheel drive steam truck, its boiler located up front, following British practice."[41]

That at least one Aultman Steam truck was made can be verified by an article in *The Automobile & Motor Review*. The information there was largely technical in nature, and confined to drawings of the vehicle, including a side elevation view of the four wheel drive principle. The boiler design for the truck is of interest here, for its basic design was indicative of the steam car's build.

It was a firetube design of extra large dimensions, boasting 32 inch inside diameter by 18 inches high. Additionally there were 1368 half-inch copper tubes, with a heating surface of 244.64 square feet, which was 15.3 square feet for each developed horsepower of this impressive boiler. The latter rating was 16, obtained from a steam engine featuring two double-acting pistons of 4.5 inch bore X 4.5 inch stroke, for 143 cubic inches. To quote the article, "An experimental vehicle was constructed and put through many severe trials, with results that have confirmed the designer's views."[42] However, in order to get a motor vehicle to market, the firm chose the car to produce first, as the most profit potential for the least investment. *The Horseless Age*, though, closes with the afterthought: "They [i.e., the company] will, however, take up the truck again soon."[43] Incidently, the Aultman was a steamer because of natural progression; its parent firm specialized in steam engines.

Previous to this, *Motor Age* revealed, "The Aultman Company. . . has constructed two steam vehicles and is now testing them. When the makers are thoroughly convinced that they have a practical vehicle[,] they will commence to make them for the market."[44] The implication appears to be that, in the interim between these two dates, apparently further experimentation ensued and the firm subsequently announced it was ready to deliver cars.

We have what amounts to a review of the Aultman Steam in *Cycle & Automobile Trade Journal*. *Standard Catalog* mentions ten as the number of cars finished by March 1901, and there is no reason to assume this is inaccurate. One of these was given a "once over" by the above cited journal. With top, the car cost fully $800, although $50 was cut from this price for models without a top. A better-trimmed surrey listed at $900. The power plant was a single cylinder of 2.5 in. bore X 3.5 in. stroke, for just 17 cubic inches. No hp rating was listed by the company or by the journal. A twenty-five gallon water tank was fitted, with the Kelly burner fired from a six gallon fuel tank of seamless steel. Usually gasoline could be used, while the boiler—also of seamless steel—was 16 in. X 16 in. A Kelly generator was likewise fitted. Total weight of the vehicle, *sans* passengers, was said to be 800 pounds. Controls for the burner, along with air adjustments and water were from inside the car; fairly typical practice for steamers of the period.

In fact, these were all accessed from the seat. The automatic air pump was also included at the above prices. "An automatic check[45] connected with the water glass gives notice when the last gallon of gasoline is reached."[46] Wheels were 34 inches and fitted with either

2.5 or 3 inch tires. Nothing in the specs was more than typical, although the journal noted the car's running gear and accouterments were of the finest construction.

An odometer and headlamps were included in the quoted prices. No performance data has surfaced on this make, and the woeful lack of information available in the steamer's home town of Canton displays in how little esteem it was held.[47] One potentially promising lead did not pan out.[48] There are illustrations in the old publications alluded to, but they are of same model, same view. Only the one photo, unfortunately, has survived.

As for production, *Motor World* provides the only clue other than those already alluded to. It says, "Several experimental cars had been built, but nothing came of them." There was, also, a brief notice about Aultman & Co. going into the hands of a receiver 'last week.' Sadly, "the firm had been preparing for a year or more to embark in the automobile business."[49] Liabilities were listed as $1.5 million. Significantly, when the firm displayed its products at the World's Fair in St. Louis in 1904, no cars were shown.

Additionally, after the 1902 listing in the city directory, a check in the 1903 edition revealed a few changes. An end-paper advertisement shows cars were already gone. The Aultman Co. was now listed as manufacturers/dealers of, "Threshers and engines, elevators,

conveying and power-transmission. . . road-making machinery, etc."
Automobiles presumably could be among the "etc.," but they are not
so listed in the main section.[50]   Little is known for certain.[51]

The rest of the tale is not long in the relating. Certainly, this
big firm had many more products to keep itself occupied with than
just automobiles. For instance, about the same time as work on the
Aultman Steam was progressing, the company was also engaged
in the manufacture of a traction engine. This steam-driven engine
operated in much the same fashion as the steam car, except its
cylinder was a much larger one and it used steel wheels, *without*
tires, for transportation. There were other ideas being explored by
company management. The company still had potential. But it was
to remain untapped.

A general strike of labor in 1903 helped nail down the lid on the
company's coffin. On May 9th, 1905, the bankrupt company's factory
was parceled off, and all subsequent Aultman dealings, detailed in
Bixler's history, cease to be of concern to us.

The cars had been one of the casualties. Of those few Aultman
Steam cars built, none have survived. There are surviving Thresher
engines scattered about the country, including a couple in the Henry
Ford Museum. As for the Aultman factory, it passed into the hands
of other concerns after the demise of the company.[52]

It is a testament to the size and convenience/versatility of the plant that several businesses were simultaneously able to occupy different sections of the factory in later years. In Bixler's book, we have two pictures of the plant, taken in the mid-1960's, which by then was deserted and forlorn. After this time, the mammoth structure was demolished completely. At this writing, an imposing correctional institution in its own right occupied the location where the factory once was. No trace of the old Aultman structure remained at the sight. Moreover, as Bixler touches on another problem, we are, "confronted with the fact that no records of the company are available."[53]

Second Marque.
The Gramm Steamer of Chillicothe, Ohio. (1902-1903)

Tracing the elusive history of the Gramm Steamer of Chillicothe, Ohio, has not been an easy undertaking.[54] Chillicothe was one of the early pioneer towns of the auto age; but the arrival of the car, "hastened the need for paved streets in Chillicothe, construction of which had begun in 1898 on only the main thoroughfares."[55] The place, in many ways, was a typical small sized turn-of-the-century city. In 1900, its population boasted just 12,976 persons.[56]

This corner post—indeed first state capital—was to be the home of eight cars, according to *Standard Catalog,* so it had its share of car building in that early period. In fact, "During the first decade. . . it appeared likely that Chillicothe was destined to become a major center for the automobile industry."[57] One of its pioneers, day bank clerk Benjamin A. Gramm (1872-1947),[58] according to one source, "built his first auto [there] in 1889."[59] This initial creation could not have been more than an experiment, since details are woefully lacking about such a car ever running on Chillicothe's streets.[60]

In 1890, Gramm's manager at the bank purchased an adding machine at his suggestion. This expedited the young man's workload, enough so that he could spend more time laboring on "his first love," in the relatively new field of automobiles. Gramm's tinkering was done in an old building behind his residence, apparently at 82 W. Fifth.

The hours Gramm spent in that shed were, well, "long and late."[61] In that he worked in similar fashion to Henry Ford, and coincidently they both shared the same birthday of July 30th—although Henry Ford was nine years older—Gramm has been compared to the latter. Not unfavorably, we might add.

In an article published in the now defunct *Ohio State Journal*, Gramm mentions his visions of horseless wagons was born, "while cycling over the surrounding roads." Similar to Hiram Percy Maxim's inspirations.[62] Gramm then appears to have spent the better part of a decade toying with such ideas. The amount of hours spent in the shop finally bore fruit around the turn of the century as the first "confirmed" mobile creation of Gramm's made its merry way on to Chillicothe's streets.

There is some controversy about what kind of vehicle this was. One source states this was a "horseless wagon," but Gramm's own account of this first vehicle does not say this.[63] In fact, Gramm is less than clear on this point.[64] Probably not on purpose, though.

Gramm states, rather laconically, as he piloted his new creation down the street, he veered suddenly toward a crowd of on-lookers. His boss, Mr. Renick, happened to be among the group. "I drove this horseless vehicle toward him and stopped it just as it touched his watch charm amid the shouts of warning."[65] To his credit, the banker remained stoic as his young companion *really* put his confidence to the test. Gramm offered the reason for this unexpected behavior from Mr. Renick's statement: "Ben explained to me with his drawings just how he could control and stop it and I have known him every day for 15 years."[66]

This charming tale offers a glimpse of the confidence invested in Gramm by this man, and no doubt by many others. Presumably, this first vehicle was constructed in that shed behind Gramm's residence. It must have been about this same time when Gramm realized he had to have more room for his tinkering. Especially if it would ever be more than *just* tinkering. In August, 1901, work began on the new accommodations.

In the October 28th, 1901, issue of *The Scioto Gazette*, the local newspaper ran a very brief article about Gramm's new business, which followed shortly on the heels of this foray.[67] This enterprise was located in a brand new building at 87 S. Walnut Street. Out of this location, with his banking career blissfully concluded, Gramm proceeded to create quite a stir in the growing automobile trade.

Apparently that first demonstration of horseless carriages by Gramm had convinced some potential customers of his sincerity, and orders started to come in steadily. The move to the new location was necessitated, no doubt, by the lack of space in the old quarters, so Gramm set about doing things right. "By 1901 [Gramm] was constructing and repairing automobiles at his own shop."[68] Gramm's new business, dubbed The Motor Storage & Repair Company, according to the newspaper account, "is the first fully equipped place of its kind in Southern Ohio."[69]

In fact, the reporter in question over optimistically predicted, "This plant is destined beyond any question to take a leading place in the roll of industrial enterprises." It must have been about this same time when Chillicothe got its first legitimate car, a Conrad steamer built in Buffalo, New York, and shipped in to town unassembled. The owner, Dr. W. A. Hall (1851-1914), employed Gramm to put this steamer together.[70]

Unquestionably, Gramm was the right man for the job as well. Indeed, he, "had the good doctor making calls in amazing time, [at]15 miles per hour."[71] The October newspaper article mentions Gramm's concern was, "putting up... one of the best machines which is now on the market." But, just below this, the same source speaks of actual vehicle construction in the future tense. Apparently, about the same time Gramm, with the collaboration of Orville Houser (1870-1935), a machinist of much local repute, "produced a racer said to have a speed of 15 m.p.h. on the track."[72]

A peek at the 1900 Chillicothe City Directory shows Gramm still at 82 W. Fifth, and listed now as in a partnership with Joseph L. Schilder (1869-1935), of 21 E. Fourth.[73] This partnership, a bicycle dealership/repair shop, was located at 47 S. Walnut, not far from the Motor Vehicle Company.[74]

An article in *The Scioto Gazette* said of Gramm's activities at the garage, "he assembled vehicles which were usually delivered to buyers in packing crates."[75] Just like the above cited Hall car. With all of this going on, Gramm must have been a busy man.

If we accept all this at *prima facie* value, it would seem at first glance the so-called "Gramm Steamer" may just have been one of these out-of-town cars assembled by Gramm and company. Most fortunately, we have been able to establish actual production.[76]

Fortunately, enough documentation survives to piece the puzzle together. One source says, "an 'electric wagon' was built about 1902," but this can not be confirmed from any other source.[77] The mere fact this reference was in parentheses should make one suspicious.

However, Gramm's interests were not confined to just one particular kind of motive power. A picture from 1902 shows Gramm's first production from the new business: the caption itself and the vehicle betray it as a truck.[78] It is just possible this was the "electric wagon" referred to above.

Obviously, this was not the only motor vehicle to emerge from the shop: Brown makes a pointed reference to Gramm's advertising of, "'rental cars for all occasions.'"[79] The October article refers to this as an 'assembled car.' "These carriages. . . being composed as they will be of selections made of the best parts of the different makes."[80] There is more.

An article in *Cycle & Automobile Trade Journal* pointed out that Gramm, "has commenced the manufacture of motor vehicles at Chillicothe, Ohio." Also, "they put out a steam carriage, but are prepared to make either gasoline or electric to order."[81] This article was dated in January 1902. Considering the lead times for periodicals in those days, manufacture of the car must have started in late November/early December of 1901. This would have been just after the newspaper article had appeared at the end of October 1901. Recall that the cited article had spoken of vehicle construction in the future tense. Obviously, since a B. A. Gramm is listed as the GM, the source of the article must have originated with him.[82]

It is patently clear that Mr. Gramm provided most of the impetus to get this car into production. After all, the car was named after him. However, Gramm apparently had partners who put up at least part of the money he needed to start his car building business.[83] Said the article, "Mr. Frank Ramsdell [sic] and Mr. Robert McVickers have the superintendence of the factory. . . [while] Schilder and Gramm have the sales department under their charge."[84]

Additionally, the Ross County Historical Society's (RCHS) newsletter from October 2000 displays a photo, dated 1903, of Gramm testing a nearly complete car. This vehicle was visually very dissimilar to that machine in the 1902 picture; it was much longer, closer to the ground, with a much more generous steering gear. Obviously, this particular vehicle was a car. Considering the lateness of the date, this must be either a Gramm Steamer or a Buckeye.[85]

According to *Standard Catalog*, the Buckeye was a four cylinder car, with the engine located up front. The chassis picture has no engine up front, but midships there appears to be a compact steam engine. At least, it does not resemble an electric motor or a gasoline engine. The car's occupants are obviously getting ready for a test run. So this must be a Gramm Steamer in prototype form.

Gramm's own account of the vehicles he built in 1902 indicated they had a double-opposed motor with a six-passenger body.[86] As to quantities, "During 1902 and in the fall we had completed six motor cars."[87]

Little has come to light about this steamer. There is the short entry in *Standard Catalog*, but few more bits of information.[88] A check with the NAHC revealed no sales catalogs or repair manuals in the collection on the Gramm. No photographs turned up, either.

As for *The Scioto Gazette*, it makes a couple of statements that could use some enlargement. "Their plant. . . [has] innovations and improvements for the manufacturing of automobiles, which have not yet been thought of by some of the [Eastern] plants."[89] Unfortunately, the journal writers stops short of telling exactly what the improvements were. Gramm's own sketchy account of those early days is confined for the most part to his much greater contributions to the truck building industry.

Like the Aultman's builders, the Gramm concern had many more irons in the fire than just the steamer. In fact, Gramm the man was busy working on quite a number of things, all about the same time.

Still, there were three models as finally announced of this steamer. A two-seater sold for $750, while a dos-a-dos fetched $850, and a delivery wagon topped the list at $900. It seems a safe assumption the power plant was designed in-house by Gramm and his associates. Unfortunately, this writer was unable to locate any illustrations of the engine or the boiler/engine relationship in the chassis. But the description as contained in the *Cycle* article bears some mention. "The feed water is run through a copper coil in the exhaust muffler." The effect of this was to make the water, "almost at boiling point when it reaches the boiler."[90] Moreover, Gramm's steamer, unlike the contemporary Stanley Steamer, utilized a condenser to bring the steam back to the tank, so it was a self-contained unit.[91] However, just like other steamers of the era, the car required much preparation and its pilot light startup-to-running time was on the order of, say, half an hour to 45 minutes.

One must surmise that Gramm changed his mind about constructing steamers almost immediately after their announcement. The preponderance of the available evidence strongly suggests just a very limited production of these cars. The reporter from *Cycle & Automobile Trade Journal* saw actual steam cars in production. That much is clear. However, none of these cars have survived. Then, for some reason, Gramm switched to gasoline engines as his sole motive source of power. Thus the appearance of the Buckeye. He likely believed the internal combustion vehicle was the best course to proceed by. Unquestionably, the prevailing business climate suggested such a course. By 1903, the gas car was starting to assert itself in the marketplace.

Do not forget, Gramm was still selling bicycles in his partnership with Schilder, even adding a line of plumbing and electrical parts to the mix. This was in addition to the still functioning garage. The latter advertised it had room for twelve horseless carriages to be stored and cared for, "at the constant service of the owners, for much less than the cost of keeping a horse, either at a livery stable or in the barn of its owner."[92]

But Mr. Gramm, destined to be a big name in trucks, soon after moved into more expanded quarters at East Second Street to build the Logan car, the third of his cars. The building at 87 S. Walnut passed into the hands of others and was at this writing a print shop.[93]

Benjamin Gramm would go on to play the key role in preparing American Liberty truck operations during World War One.[94] As for the Gramm Steamer, it just passed into history. Instead of a vibrant, going concern, this car became an abandoned, almost forgotten symbol of a lost time. Much like the steam car in general. For his numerous contributions in the early days of the automotive industry, Gramm, in 1946, was invited to participate in the Golden Jubilee celebration in Detroit.[95]

Third Marque
The Krotz Electric of Springfield, Ohio.(1902-1905)
The Krotz Gas-Electric of Defiance, Ohio.(1908-1911)

Alvaro S. Krotz (1864-1954) was born in a log cabin near Defiance, Ohio.[96] From these rustic, quaint beginnings, he would rise to become a major figure in "horseless transportation." For this early pioneer, the "auto bug" struck early and often. In fact, he was to be involved in the design and fabrication of automobiles for decades.[97] By 1903, Alvaro had long since migrated to Springfield, Ohio.

The Krotz Electric of Springfield, Ohio.

A car identified as "owned by A. S. Krotz of Springfield, Ohio," was featured in an advertisement for the Willard Storage Battery of 49 Wood Street, Cleveland, Ohio.[98] The caption is very clear; it states the car, "has been operated for over three and one half years on the same battery and is still giving full mileage."[99]The accompanying detailed drawing is of a rather ordinary looking electric car. Apparently, although its appearance (in common with many of the early cars) left something to be desired, it was one tough car. We are told, by Krotz himself, he built this vehicle and he unquestionably did.[100] This was impressive enough, by itself. His automotive roots ran much deeper than that, though. Like so many of America's early auto pioneers, Krotz started tinkering very early in life.

The first instance on record of Krotz actually *building* cars of any kind was back in 1878 when, "he built a small engine and put it on wheels." When we consider he turned just 14 that year, this was noteworthy. Further, "it was similar to the automobiles in use at the present time."[101] This "present time" reference was from around the turn of the twentieth century. Alvaro's skills evidently were not confined to automobiles either, for when he moved to Springfield in 1890, "he designed the [city's] streetcar system which. . . has been such an important factor in facilitating the business of the city."[102] He did all of this by the ripe "old" age of 26!

Krotz's early start in the world of work was not, of course, strictly automotive or even transportation related.

It was not long, however, before Krotz became more acquainted with companies that naturally tended him towards this still new field of cars. This appeared to be his field of choice anyhow. Nor was higher education his only tool. With a background from living and working around on the farm while at home, Krotz soon became mechanically proficient. In 1897, he became associated with the Kelly Rubber Tire Company[103], and, in 1900, the Grant Axle & Wheel Company.[104]

In fact, by this time he had already had an eventful couple of years building cars. When Krotz joined up with the Grant outfit, one of the first things he did was to construct a number of nearly complete chassis—according to one source, 25 units—for the Pierce Electric Company of Bound Brook, New Jersey.[105] There were other customers as well. The degree of completion of the chassis is not known, but certainly this gave Krotz valuable experience in the business.[106] Nevertheless, this was still going to be a sideline business. With Krotz effectually working for others, this would remain a sideline in Springfield.

Not that the business climate in town was unfavorable. Thirteen different marques, as of this writing, have called Springfield home. Several saw production, and the local populace must have welcomed Springfield-built cars with some satisfaction. Several of the other marques in town were being developed just about simultaneously as Krotz was so occupied. At least his experiments did not exist in a vacuum. But, unfortunately, details are lacking.

We *do* know, in each of his early settings, Krotz had managerial oversight, but he was still working with his hands. When the Grant firm suffered a devastating fire, Alvaro Krotz used the unexpected opportunity to turn to working full-time on his own inventions. He was already doing this on a part-time basis.

His various activities soon made him into a respected, well-known businessman.So when the word was passed Mr. Krotz had been using an electric car and getting extra long life from the battery power, people far and wide paid attention.[107] The adoption by this seasoned entrepreneur of a motor car (still new enough to warrant public attention) for his personal transportation was significant. Still more when it was surmised he was engaged not just in driving—but also in building—them.

Of course, the active assembly of cars predated this time. By 1902, Krotz was a pro; he was already, "engaged in the manufacture of automobiles[,] and quite a number of electric automobiles have been built under his direction."[108] We are also informed of the company's great success, with advance orders already earmarked for customers before completion.

The 1902 Springfield City Directory merely confirms Krotz was the Superintendent of the Grant outfit.[109] Data are limited. For instance, we do not know if Krotz were engaged in motorcar manufacturing on his own, or for the Grant outfit. He was not yet listed as an automobile manufacturer in the directory, although he was well-established. There is some controversy here. One source indicates Krotz started building cars in 1903.[110] But, in yet another extraction, Krotz Mfg.[Manufacturing] Company is listed as making the Krotz Electric from 1902-1905, in Springfield, Ohio.[111]

All of this argument notwithstanding, it appears at least part of Krotz' plan for making cars involved perfecting an electrical storage battery of his own design. It seems logical the inventor was interested primarily in a satisfactory venue for this battery. Allowing for advertising hoopla, if only half of the Willard Battery claim[112] were true, it would still be impressive achievement indeed. And certainly better range than the typical contemporary electric could have afforded.[113]

Therein lay the problem. The chances for Krotz to provide a better design electric did not look any better than for other inventors of the time. Electric power for homes was an important step in modernization. But doing the same for cars was proving quite troublesome in the early twentieth century.

Evidence for Krotz's experiments in and of themselves is scarce. However, there is more to the paper trail. The 1903 directory lists Krotz' company as the s.e.c. (Southeast corner) of Linden Avenue and Monroe Street.[114] By this point, he had obviously gone out on his own, as we have seen. The nature of his new business was patently mechanical, with greater emphasis on automobiles, although the only listing in that year's city directory of an auto manufacturer was of Emil Koeb of Koeb-Thompson fame.[115]

One of the contemporary automobile magazines elaborated in this time period on the existence of the Krotz firm. "A. S. Krotz... has been experimenting for some time with electric vehicles and is now about ready to market machines."[116]   Mentioned once again in this notice was the fact the cars would be, "equipped with a condensed storage battery of his [Krotz's] own design."[117] In line with this, the city directory lists "machinery" as Alvaro's business.[118]

Few details about this Springfield car are extant, except for our elaborate drawing. From this, we can see a largely conventional high wheel design,[119] complete with wooden spokes and large (apparently) solid tires. The single enclosed electric motor resided sideways under the two-passenger compartment. Drive was by belt to the rear wheels via pulley, and the driver used a typical tiller for steering. For suspension, the carriage used unusually heavy-duty springs, which presumably did not provide much cushion from the roads for those poor pioneer enthusiasts. Little of the construction seemed prone to inspire imitation, but it was also likely of solid build. The enclosed battery compartment was behind the seat, which served to protect the occupants[120] from the obvious physical drawbacks associated with batteries. This was common to electric cars. We can safely surmise that much of the vehicle weight was taken up with the batteries. This also really intruded into the available passenger space. At least on the electric cars, there was usually no gear shifter protruding from the floor or on the outside corner interfering with the available passenger room. The Krotz car lacked a gearshift.

Now this heavy, ponderous electric could not have sped along at more than a stately pace. This fact, in retrospect, was really a good circumstance, with the fully exposed passenger compartment and all. Nonetheless, early daring, long-suffering motorists mostly did not give this matter a second thought. There was a basis for this. Krotz's design appeared to be worthy of trust. Nor was the motive power alone of concern to the inventor. Krotz's capabilities as an innovator apparently knew few bounds.

In addition to his mechanical conveyances, Alvaro took a brief foray at tire design. He made himself some additional renown—not to mention some cash!—with a new, rather novel idea for a notched tire. This had a feature which would appear odd, at first glance, to latter-day readers: the tire had tread. We must understand the rubber tire was a fairly new innovation.

As such, and in common with so many new inventions, erroneous ideas often coexist for a time. There were a host of such beliefs apparently associated with this tire. For instance, some people believed cutting away some of the tire would weaken it. Krotz reported a few folk actually bit into the rubber.[121] But the early business concerns in the field of tire manufacture, at least the ones which had enough insight, recognized immediately they had a winner. The Kelly-Springfield firm promptly bought the rights to the design.[122] The word quickly got around.[123]

While all of this was unfolding, the erstwhile car maker was turning his attention to what motive power his future cars might utilize. The electric storage battery and the consequent limiting factors of the electric car gave Krotz time for pause. By this point, it was apparent that the electric as a power source for cars had serious drawbacks.

Krotz was no starry-eyed dreamer. The inventor had to be aware of declining sales for electrics, and some headaches that might not find a solution anytime soon. All of this would inevitably be a restricting factor in the further physical development of electric cars. Not to mention their marketability. As a result, Krotz early on started looking to incorporate the internal combustion engines into his future designs.[124]

This was about 1905-1906. Krotz promptly worked up the design for a high wheeled gasoline vehicle which had a great deal of practicality for the day. It also happened to be a better looking product than he had heretofore produced. The most noteworthy continuing feature was the high wheels; very appropriate in view of the often poor, deeply rutted roads early motorists might encounter. The power source, of course, had been changed to one that might have a better future. Krotz was wrestling with this whole power source issue, and it had to be still resolved to his satisfaction.

Meanwhile, the company itself was moving again. This time, with the manufacture of automobiles firmly established, the new quarters were at 69 N. Center Street.[125] Ironically, this also happened to be the last year Krotz was listed as an automobile manufacturer in Springfield.[126] No doubt by then his vision was turning to bigger and better things. Like the high wheel gas buggy he hoped would make it big. And, for some reason, someplace other than Springfield.

Likely, though, the reason was financial. Krotz needed some ready monies for working capital.

With this in mind, Krotz removed himself to Chicago and promptly cast about for a new 'junior' partner. Had the resourceful businessman possessed sufficient finances, he would likely have manufactured this new car himself. Krotz instead found a willing partner in the form of Sears, Roebuck & Co. to do the marketing for him.[127] The latter firm was just then looking for ways to expand its already famous catalog line. So it listed the high wheeler, and bankrolled Krotz to set up a factory for the car. Soon the factory at Harrison and Loomis streets in Chicago was cranking out finished cars. Alvaro Krotz was the factory supervisor, overseeing the production of the car.[128]

Yet Krotz remained dissatisfied. His fresh high wheeler was just a modest internal combustion car devoid of amenities. Thus Krotz had given up on a pure electric; it did not seem to be his forte. And the straight gasoline car turned out to be less than he wanted. His next logical step then was to take the best features of the two types. The wily inventor showed his latest proposal to Sears supervisors. They thought the whole idea too impractical. The powers that be may not have been very enamored about the high wheeler's success either.[129]

The Krotz Gas-Electric of Defiance, Ohio.

So with this avenue closed, Krotz subsequently returned to Ohio to raise capital to build this latest car. Wanting to avoid the earlier Springfield debacle, he migrated to his hometown of Defiance to secure financial backing. In addition to ready monies, he also needed a secure sight for a factory location.

Few details of the means by which Krotz sought his ends are extant, except we learn: "Mr. Krotz has been quietly at work in this city for the past months and as a result twenty-five local men have become interested in the new company."[130] Some of the more prominent of these "twenty-five" read like a *Who's Who* of some of the city's most important individual businessmen.[131]

Krotz's hometown was no stranger to automobiles and their construction. In fact, Defiance was to be the home of three cars,[132] including one called the Highway, which was announced in 1920. Its parent firm was the Golden, Belknap & Schwartz Company. A small world for this firm counted several influential industry leaders like Charles Kettering and other prominent individuals under its heading. Nevertheless, despite an impressive pedigree, not a single car appears to have been produced.[133] A second machine, called the Defiance, made by the Miller Machine Company, was advertised for sale in 1909. Despite a decent buildup, nothing much seems to have come from this endeavor either.[134] The Krotz Gas-Electric, in this respect, was unique. It *did* see production. That much is well established.

Before there could be a new car, though, a company was needed. Early in February, 1908, following what must have been brief negotiations, the Krotz Autobuggy Company was indeed incorporated. According to *The Horseless Age*, "The company will erect a factory in Defiance and manufacture a gasoline-electric buggy."[135]   Undoubtedly, Krotz utilized whatever funds he had made from the Sears adventure as well as other monies from the local investors for this new firm. *The Horseless Age* was not alone in keeping itself abreast of the latest developments. There were other publications taking notice, too.[136] The dual power of the vehicle must have been attractive to the national press. Not to mention this healthy curiosity about what Krotz might come up with next after his days with the Sears car were through.

No doubt much of the *local* interest was due to the timing of the new car and its impact on the economy. On a narrow view, a new business meant new jobs and more growth. Unlike the enterprises in Springfield, for which cars were always just a supplement to the existing business, the entire effort of the Defiance concern would be automobiles (and light-duty trucks) from the beginning.

In fact, according to one account, "Defiance is landing a manufacturing industry during the period of business depression [being] felt over the country. . .."[137]   Unfortunately, many details of this latest enterprise are lacking, and we know little of the everyday workings of the factory. Nor do we know the actual influence Krotz's investors had on the daily output of the factory.

However, we do know the kinds of vehicles to be offered. According to a contemporary reference, the new "auto-buggy" will be marketed at a price, "that will appeal to the farm trade and others desirous of a cheap machine."[138] Nor was a car the only carriage to be offered. The local newspaper said the new firm would, "be turning out auto buggies and wagons."[139] We can safely surmise from the above that Krotz was trying his best to market vehicles to as wide a range of the buying public as possible.

However, Krotz' new adventure must have been alluring in other respects as well. His involvement with the High Wheeler, its development and the final marketing of a saleable product must have helped demonstrate to uncertain investors what the wily inventor was capable of.

And he was not alone in this endeavor, by any means. One of the investors included, besides the above named individuals, a D[aniel] F[remont] Krotz.[140] This brother to Alvaro undoubtedly chipped in as much as he could to help make the new car a reality.[141]

The location of the Krotz factory was Fort Street between Clinton and Wayne Streets.[142] This location was right along the waterfront, and was especially built to manufacture the car.[143] We do know several sights were carefully considered before that particular location was chosen. Spots that, for one reason or another, were rejected. No doubt, by a consensus of Krotz and of his investors.[144]

The dual power of the vehicles was clearly spelled out in the local newspaper. It stated, "their motive power will be either gasoline or electricity."[145] This would seem to imply the motive power was to be exclusively gasoline engine or electric motor. Or even that the operator of a Krotz Gas-Electric could switch back and forth between the two power sources, if he so desired. It would appear the inventor was trying to address the whole weight issue as well simultaneously. In words that no doubt echoed what Krotz himself undoubtedly uttered, his cars, "will be built light but durable."[146]

We have already noted the low price, although just what the prices were was not available at this writing.

There are few deatils other than the above available. The factory cranked out vehicles in limited numbers until 1911. That was the year identified as the end of production in the 1938 newspaper article.[147] The Sanborn Insurance Map identified the 'electric motor' business at the sight of the factory in 1911, as we have noted. This was no doubt a closely related field that Krotz took up after his car was discontinued.

There were no repair manuals or literature surviving on this car at this writing. Whatever the production was, it must have been small. It follows sales would likely have been mainly limited to the local area. None of the cars have endured to the present.

As for Krotz himself, he moved on to become a patent lawyer for his livelihood. He continued tinkering for the rest of his life. In 1946, he was invited to become a member of the Automotive Golden Jubilee Honorary Pioneers' Committee. Both he and Benjamin Gramm (from our Gramm Steamer of Chillicothe) were among 36 men so esteemed.[148] There is some justice in that.

Fourth Marque
The MacDonald Steamer of Garfield, Ohio. (1923-1924)

The newest of the marques we will look at in this thesis was also the only one not located in an Ohio city. In fact, Garfield, Ohio—not to be confused with Garfield Heights in Cleveland—in the 1920's was just a little hamlet, situated just one mile north of Damascus on Route 534 in Mahoning County.[149]

There, in a scenic stretch of rural northern Ohio, was a curious place of choice for a car plant.

Nevertheless, the MacDonald factory was located there within easy access of a busy railroad line, just east of Garfield. In the words of the builder himself, "a substantial shop and office building has been erected and is being equipped with modern production equipment."[150] There must have been considerable expense involved in locating and outfitting at this sight, since practically everything the MacDonald firm would require had to be shipped in to a relatively isolated location.

This MacDonald (sometimes spelled 'McDonald')[151] was built in the early twenties, just as the last wave of steam car resurgence to date was sweeping across America. Reason enough for including the car with the earlier alternative cars.

The builder of the car was one Duncan MacDonald, a rather flamboyant character who had previously been associated with the Gearless Company of Pittsburgh, Pa.[152] Before this gentleman became "involved" in activities that would send him and some business associates to prison, though, he would have some surprising successes. We can follow this checkered career closely.

For example, it may have nothing to do with the situation, but there is a good paper trail on this newest of our alternative makes. As a result, we can even trace this car from the planning stages on. In a 1920 article from *AutomobileTrade Journal,* dated May 15th from Alliance, Ohio, the reporter stated that, "Pittsburgh capital" had purchased a 36-acre plot of land at nearby Garfield where, "a factory will be built for manufacturing gearless automobiles."[153] The MacDonald slowly evolved alongside the Gearless car from nearby Pittsburgh. As it worked out, the two cars bear a passing resemblance to each other, but they were not identical by any means.[154]

Both of these cars appeared in the steamers' twilight years. As such, the vantage point of several decades removed from the fact makes it difficult to see how the two makes, indeed all remaining steamers, were viewed by contemporary society.[155] Fortunately, for our purposes, we had some living eyewitnesses to converse with. At least about the McDonald Steamer. This was simply not possible with our earlier marques.

The above report from *Automobile Trade* cited "unlimited capital" behind the MacDonald. The source of these funds was likely derived from stock Duncan and his associates sold for the Gearless Company. All of this was to the tune of some $1,360,000.[156] Five machines had already been tested by then, again, according to our source.[157]

The contemporary MacDonald was most often seen, when it was that is, in the form of a sketch, which Mr. Marvin indicated was drawn by a, "third grade pupil."[158] This rendering, which, "was a poor one and out of proportion, "[159] was of a 1923 McDonald "Bob-Cat" roadster.

Fortunately, better materials were forthcoming for our purposes. The writer, through the kindness of Mr. William Lewton (1921-   ) of Lisbon, Ohio, obtained what appear to be original engineering renderings for both the aforementioned model and a sedan version.[160] The former was identified in the drawing as a Model 3-28.

The accompanying illustration is clearly of a three cylinder vertically-placed steam engine, mounted up front and integrally built with a rectangular boiler using pressure type 16 gage Shelby Steel tubing.[161]   The burner was of the atomizing kind, and the engine boasted dimensions of 3 in. bore X 4 in. stroke, for a total of 85 cubic inches of swept volume. Superheated steam was produced at about 450 psi. The whole arrangement fired by kerosene. This power plant was apparently of MacDonald's own design and featured poppet type valves removable with the cylinder head. From all indications, it was designed with easy maintenance in mind. Among the details, the steam was ignited by a spark plug, a design feature borrowed from gasoline engines.

There was still another power arrangement for the Steam Bob-Cat as well. Mr. Lewton provided a copy of an original sales brochure for the MacDonald car.[162] This item clearly shows the boiler still mounted up front, but the engine itself has been placed at the rear of the car, in combination with the rear axle.[163]

Since none of these documents bear a date, it is hard to determine in what order the cars appeared. But it does seem the 3 cylinder car was a first series, and the second, rear mounted engine was the second, and final, release.[164] Mr. Lewton also elaborated on a previously undiscovered difficulty with the latter engine.[165]

The Steam Bob-Cat was a sporty, nimble roadster, riding a 106-inch wheelbase. This model was available ostensibly in either Brewster

Green or black, with the three passenger compartment "finished in Spanish Leatherett." Again, according to the Specifications, the firm planned to market the car for a list price between $750 and $1000. The Prospectus accompanying the engineering drawing indicated the Steam Bob-Cat, "now being created is a powerful[,] classy steam SPORT CAR." It was designed as an alternative to, "any of the heavy Steam Cars now offered the American buyers."[166] Whatever market remained, that is. By this late date, as we have observed, there were very few steamers of *any* kind on the market.

There is another complication. The Prospectus bears no date, so there is no way to know *exactly* when it was prepared. Mr. Lewton had in his possession what must have been an earlier version of this information, in a more incomplete form.[167]

At least one of these cars was built.[168] This occurred soon after Duncan MacDonald left the Gearless enterprise in 1920 to devote his talents wholeheartedly to the steamers in Garfield. To quote the Prospectus again, "Mr. McDonald. . . packed bag and baggage, disposed of dictating stock interest and staked his fortune on the future for the Steam Bob-Cat."[169]

Also noteworthy was the manner of MacDonald's departure from Pittsburgh to more rural surroundings in Ohio. Duncan apparently brought few ranking company men with him. A comparison of

Gearless Company officials with the Garfield entity,[170] if nothing else, certainly showed nepotism was alive and well in the Roaring Twenties.

For reasons unknown, in fact, none of the senior Geerless officials made the trip to Garfield. Duncan MacDonald was, of course, the new company's president and general manager, with an F. N. MacDonald installed as vice-president and secretary, and an E. M. MacDonald listed as treasurer. The manager of sales was one H. H. Lang. Perhaps another relative of some kind.

These gentlemen had quite a task on their hands. Starting any new automobile manufacture was already by then highly risky. For one that specialized in an alternative car type this was even more so.

To make matters worse, there was also an economic recession in the United States around the same time. Coupling all of these factors together must have cast a dark shadow over the whole enterprise. But MacDonald remained "cautiously" optimistic. He reasoned, without a hint of justification, this recession, "should save for the MacDonald Steam Corporation thousands of dollars while forcing large production."[171] Apparently MacDonald believed mills and factories would be forced to lower prices and offer other incentives to attract his business. Such as it was.

Now, we must admit, Mr. MacDonald was something of the idealist, it would appear. But he was also one determined individual. One not easily dissuaded.

Nevertheless, a long gestation period ensued before the MacDonald enterprise finally bore fruit. There were bound to be local reports as progress was made towards production. A short article appeared as a front page piece in *The Salem News,* a local daily newspaper, in late 1921. One working car was reported to be having demonstrations of its prowess every Friday and Saturday nights in Garfield; these trials were, "proving very satisfactory."[172] Nonetheless, national publications also chided in, albeit at a much later date. *The Automobile* in May, 1923, a full year and a half after, reported car production was to start "soon." In a Press release dated May 7 from Alliance, Ohio, the writer noted, "production is soon to be started on steam automobiles."[173]

Despite this "low" production, the firm did not eschew advertising. It touted the new Bob-Cat was, "flexible to a degree never possible in a gasoline-propelled automobile." Additionally, there were, "none of the inconveniences of the gasoline motor; nothing but a smooth flow of power whenever you want it."[174] Notably, the last of the Stanleys had similarly worded ads. Problem was, the Stanley was a known, and established builder with a quality reputation; the MacDonald enjoyed none of these advantages. Besides, the Stanley firm by the twenties had troubles aplenty of its own.

Nonetheless, the MacDonald company made a valiant attempt to sell the public on the value of steam. The same ad mentioned steam was, "the source of power that many experts contend will ultimately replace gasoline."[175] Just who these experts were and why they believed that way was not disclosed in the ad. Most likely, the experts were the very same people employed at Garfield at the factory.

Mr. Lewton recalled that, in his youth, he was acquainted with some of the men who worked there. One of those was a Mr. McCoy, an electrical engineer or a machinist. There were others, but he could not remember their names. "They worked many a long hour at the plant; preparing the car for showings and that kind of thing," recalled Mr. Lewton.[176] This was natural. The workers wanted to present their machine in the most favorable light possible.

Obviously, some thought had gone into the car's advertisement. The vehicle's best features were highlighted, although the car rendering was still no more than a hand drawn one. At the bottom of the same ad, "responsible dealers" were urged to carefully look over the products of the firm. Their inquiry would be, "handled by the head of this organization."[177] Except he would have other, more pressing matters to attend to by then.

There is an eyewitness to the events mentioned in *The Salem News*. The writer interviewed Mr. Earl Mather (1919-   ) of Garfield, who was born and grew up just down the street from the factory.[178] Mr. Mather recalled one vivid childhood memory of the MacDonald.

"One evening [men] came down the road with a vehicle. It was just the frame of a car. It did not make noise, so it had to be a steamer. I remember it was really silent; so much so, you could hear the men aboard talking clearly. There was a box for one man to sit at the steering wheel. The others weren't [sic] sitting. It appeared they all had jobs to do to keep the thing running. They all went down the road towards Damascus, on the thing, all without a sound. It was almost dark when they got back."[179]

Mr. Mather stated he never saw complete cars emerge from the factory. He could not recall much more of the foregoing incident, as he was not yet five at the time. Nevertheless, given the interval of the passing years, the impression upon the lad must have been lasting.[180]

When the writer inquired about the factory, Mr. Mather quietly pointed out its location. He also stated after the plant went idle, it was boarded up. This was probably done at the instigation of the proper authorities. The facility started to run down after that, and eventually fell down.

Mr. Mather recalled going into the factory as a young boy soon after its closure. He said he was surprised by the number of wooden patterns he found lying all around, along with a host of automobile bodies in the then deserted building. He further remarked he did not believe the assemblies he found would have been useful for steamer constructions. Mr. MacDonald apparently had purchased new Chevrolet bodies; he was obviously planning on modifying them by adding additional lengths at the front to accommodate the steam engine.

Mr. Mather personally did see the bare bodies, but thought they were shipped in. He did not recall any specific details about car bodies actually being made in Garfield. However, Mr. Mather personally saw a partially completed car with a Mohair dashboard and a glove box.

A William Meiter was another potential eyewitness to actual tests of the cars. According to *The Damascus Herald*, this gentleman saw one of the cars stuck in the mud. "He could see the wheels spin[,] but could not hear the engine running until he got close to the auto."[181]

Like the other little known marques, there has been some controversy about key elements of the MacDonald's story. Two sources list total production as about "48 cars."[182] *Standard Catalog* stated that, "1924 output of 1,000 units would include touring cars, sedans[,] and coupes."[183]

This was Duncan MacDonald's spiel; whether it was just an idle boast or the businessman actually intended to follow through is not known. For certain, however, MacDonald's efforts were not confined to just complete cars.

The firm also, "sold conversion kits for internal-combustion engines."[184] *Standard Catalog* indicates, in fact, that, "production [in Garfield]. . . had been almost exclusively devoted to [those] conversion kits."[185] The problem is, no illustrations and little other information had survived to this writing on these kits, other than our previous reference.

At this writing, the "remains" of the factory were on private property well off the main road. There is the concrete slab, with shrubs and bushes shooting up through the cracking surface. As Mr. Mather said, "unless you knew what you were looking for, you wouldn't [sic] see it."[186] Finally, Mr. Mather sighed and said he thought personally Mr. MacDonald meant well; he just ran out of money. Mr. Mather emphatically insisted this man was not a crook. "People had bought stock in the company, then just lost faith in MacDonald. They just didn't [sic] have patience. He needed time. They just didn't give him any."[187]

Whether MacDonald did have good intentions will never be known, for he had more pressing matters to attend to. In May, 1923 MacDonald was indicted by a federal Grand Jury, charged with Conspiracy and Mail Fraud.[188] Three other former Gearless officials were charged at the same time. In January, 1924, MacDonald and two of his compatriots were found guilty as charged; all three were then incarcerated for a time.[189] One more thing was certain: MacDonald's conviction meant the end for his beloved steamer in Garfield.

The doors of the factory were summarily shut. Mr. Lewton sadly recalled the factory was dismantled and sold during the Depression at Third Street in Salem just for the price of the scrap. He also informed the writer that, following the closure of the MacDonald firm, the one completed sedan in the building at the time was confiscated by the Federal government and put into a nearby storage building at E. L. Grate & Co. in Salem. This firm happened to be the town's Willys-Overland car dealer at the time.[190]

Mr. Stanley recalled seeing a MacDonald car infrequently about town in later years, often in relation to some special occasion or other. That was the only car of this make he ever remembered.[191] Mr. Lewton recollected the storage building and its contents burned around 1948. What was left of the car after that was subsequently scrapped.[192] The disposition of the Steam Bob-Cat is unknown; total production of either model is also unknown. But could not have been more than a few cars. As John Bentley noted, "manufacture does not seem to have progressed much beyond some prototypes."[193]

In retrospect, Mr. MacDonald must have known when he moved to Garfield he was already in serious legal difficulties, but he never wavered in his resolve. Moreover, following his release from prison, Duncan once again became involved in automobiles, specifically steam cars. In 1930, he collaborated with another daring businessman named Jeffrey Carqueville on modifying a Nash sport touring car to equip it with more "modern" steam equipment.

Quite possibly, this was a variation on one of Duncan's portable conversion kits. The flash boiler, which could handle 800-1000 pounds of pressure, was mounted under the hood. The engine was a four cylinder, mounted at the back, directly driving the wheels without gears. Unfortunately, this idea was stillborn, as the two businessmen could find no serious investors in a steam car enterprise by that point.[194] No matter how good it was.

If nothing else, this postscript showed MacDonald's continuing enthusiasm with steamers. He may have been more sincere about building cars than he has been given credit for. Back at Garfield, he had continued development on his MacDonald car even when he knew he probably would not get far with it. Why is hard to explain. The most plausible explanation is MacDonald, once he realized the Pittsburgh car was a scam, left Geerless with every intention of manufacturing cars in Garfield.[195] The on-going experiments and very frequent car demonstrations, some as far distant as Youngstown, seem to suggest more than just *another* scam. At about the same time, the firm was sending out the prospectus to get potential dealers involved in the MacDonald car, without missing a beat.

Fifth Marque
The Thresher Electric of Dayton, Ohio. (1900)

At the turn of the twentieth century, Dayton was an essential cog in America's industrial expansion. Now the industrial base of this part of the United States eroded in the intervening century, we must admit. Despite that fact, at this writing, the city remained one of Ohio's most vital medium-sized cities.[196] In the last century, many concerns have come and gone there. Among those, the Thresher Electric Company was just one of two entities in Dayton bearing the name of Thresher from around the turn of the century.[197]

The Thresher Electric Company appeared in the 1893 Dayton City Directory, as the Shawnhayn-Thresher Electric Company, located at N. Library Lane between Main and Jefferson Streets.[198] This initial address was but temporary, for by the time of the 1900-1901 directory, the firm's moniker was simply The Thresher Electric Company, its location by then at an elusive entity called the Callahan Power Block.[199]

Considering the time frame, it was almost inevitable that a firm listed as a manufacturer of "electric machinery" in contemporary city directories should proceed into this brand-new field of automobile manufacture. Or at least experiment in that general direction.

The story of the Thresher Electric car is both short and informative. The 1899-1900 Dayton directory lists the manufacturer as one Alfred A. Thresher, who resided at 127 W. Monument Avenue.[200] In that year's business section, Thresher Electric was the only concern listed under this heading of "Electric Machinery."[201] Moreover, this car was truly a pioneer in Dayton. As it was built in 1900, it predates most automotive-building activity in Dayton.

Nonetheless, looking down the list of electrical-related companies operating in Dayton around the turn of the century aptly demonstrates at least part of the reason why Charles Kettering chose to start up (no pun intended!) his DELCO (Dayton Engineering Laboratories Company) concern there in Dayton about a decade after. The double advantages of the pool of local talent and the atmosphere of the city's electrical heritage must have been alluring. The greatest contribution from this concern was to be the Kettering Self-Starter for gasoline cars; which we will get to by and by.

As for the Thresher Electric car, there are a couple of features associated with it that are beyond dispute. First, since its parent firm was a manufacturer of electrical components, one can safely surmise the firm built its own components for the car. In fact, the car was apparently built entirely within house, with the exception of the bodies and, possibly, of the tires.

In addition, the machine itself was of rugged construction. It was built, according to one reviewer, because of, "the increasing demand for an electrical vehicle that. . . in efficiency, durability[,] and ease of manipulation [would] meet the severe requirements demanded of the automobile."[202] Another road test for the car appeared in *Motor Vehicle Review*. According to that periodical, "it [the Thresher Electric car] has been put through the most severe tests. . . and the results have far surpassed the most sanguine expectations of its inventors."[203]

Indeed, for once, journal copy may have coincided with the realistic aspirations of its builders.[204] What these gentlemen thought of it seem far removed from the frail appearance and expectations of contemporary horseless carriages like the Locomobile; for the Thresher was immediately put through its paces. One of the two illustrations of the brake accompanying the *Motor Vehicle Review* article shows a total of eight well-dressed gentlemen surmounting the car. Some were standing, some were sitting, and one man was even reclining on what would pass for a bumper on more modern cars. Not bad for what was ostensibly a four passenger car, with dos-a-dos seating.

This was, apparently, the first test of the car, in which it was, "carrying 1,500 pounds of passenger weight."[205] The ruggedness of the body can be attributed to the builder. This work had been farmed out to the local Woodhull Carriage Works, located at 5th Street and Home Avenue.[206] Morris Woodhull produced a line of carriages which had to be among the toughest around. Both magazines reported on this feature. *The Motor Vehicle Review* opined, the Thresher body, "has the advantage over many vehicles in having the appearance of being constructed especially as a motor vehicle."[207] We must not forget, this *was* the dawn of the horseless carriage age.[208] So what were the results of that first, pioneering road test? "This abnormal load was handled with the greatest ease, there being no perceptible additional effort. . .."[209]

The four tires were of two-inch solid rubber, the so-called "cushion" variety we will look at a little closer later in the concluding section of this paper. The left side tiller steering was typical, but there was certainly a noteworthy feature. The weight of the frame was supported at three points. This was a very early attempt at independent front suspension, but was notably unlike that of the other early proponent of this type, the Marmon from Indianapolis, Indiana. The latter boasted a double three-point suspension.[210]

Whatever the design, though, the effect was to give, "great flexibility to the frame, and [allow] adjusting itself to the inequalities of the road. . .." The result was to give the front axle freedom to pivot, which of course meant, theoretically, it could compensate for prevailing road conditions. Thus, "it is possible to make sharp turns at full speed without the least tendency to tilt."[211] This was a factor obviously thought necessary by the car's builders as the roads around Dayton, like much of the interior of the United States, left much to be desired.[212] Fortunately, the Thresher seemed adequately up to the task at hand. The available pictures make clear the powerful springs used to isolate the passenger area (compartment being too strong a word here) from the rugged terrain beneath.

The advantage here was clear enough, as the typical electric, "usually had the wheels solidly fixed to it. . .."[213] In the power source department, this car deviated from established practice by employing a separate 1 ½ hp motor attached for each wheel, geared, as we have seen, independently. Each of the two motors was itself of heavier build than those manufactured by the Thresher concern for other electrical purposes. The units were entirely encased as protection from the elements, although they, "are so designed as to afford ease of access for inspection and renewal of brushes."[214] We have already observed this was truly the era of the do-it-yourself mechanic.

As for that constant big, bulky "companion" to the electric car, the specifications are no less impressive than the above. The 40-cell, 120 amp hour Fauer battery was of four bank construction, and total weight was a staggering 800 pounds. This summed up the major problem: way too much weight! A total of 2200 pounds for a relatively small car, without a hint of passenger weight or *their* needed equipment.

With a little nod towards journalistic license, one of the reviewers states, "with this extremely light equipment[,] the vehicle has been driven more than fifty miles over hilly country roads on one charge." The gentleman mentioned the lack of effort needed to motivate the car; and, "the ease with which the vehicle surmounts the steepest hills has proved its adaptability for this class of work."[215]

Although the reviewer had to admit that no sports model was being prepared, he did mention it was possible to obtain four different speeds using a, "special method of control, without introducing external resistance nor too great manipulation of the batteries."[216] Moreover, by means of this control, the vehicle could be driven in reverse as well. With the same speeds, to boot.[217]

Details of this control device have been lacking, and we have no pictures of it. The description in the reviews is also rather skimpy. However, the effect was one of precise control for the driver, so he could concentrate on driving.

There are a few problems here. The Thresher Company wholeheartedly announced its intention to go into the car manufacturing field, making, in the process, a, "complete line of vehicles," with runabouts, broughams, even all classes of light- and heavy-duty delivery vehicles included in the mix. That much is clear.

However, the writer has been unable to discover evidence of any kind of car manufacture other than the aforementioned brake. The trail goes cold from there. No figures on vehicle production have been forthcoming. Nor have brochures or repair manuals surfaced in any form. And there are no cars extant. Finally, for some as yet unexplained reason, the firm bowed out of the field soon after. "Production had ended before 1901," said one source.[218]

ILLUSTRATIONS:

1.) The only known view of the Aultman Steam car. (NAHC).

2.) The only surviving view of the four wheel drive Aultman Steam Truck. *(The Automobile & Motor Review)*.

3.) Contemporary photo of the Gramm factory in Chillicothe, Ohio from 1902. Mr. Ben Gramm is the second from the left, holding his son. (Courtesy of *The Scioto Gazette)*.

4.) The only known view of the Krotz Electric Car from Springfield, Ohio, circa 1903. (NAHC).

5-6.) The only known sales literature on the MacDonald Steamer from Garfield, Ohio, *circa* 1924. (Mr. William Lewton of Lisbon, Ohio).

7.) Contemporary photo of a Thresher Electric of Dayton from 1900 *(The Horseless Age* and *The Motor Vehicle Review)*.

The Aultman Steam Carriage.

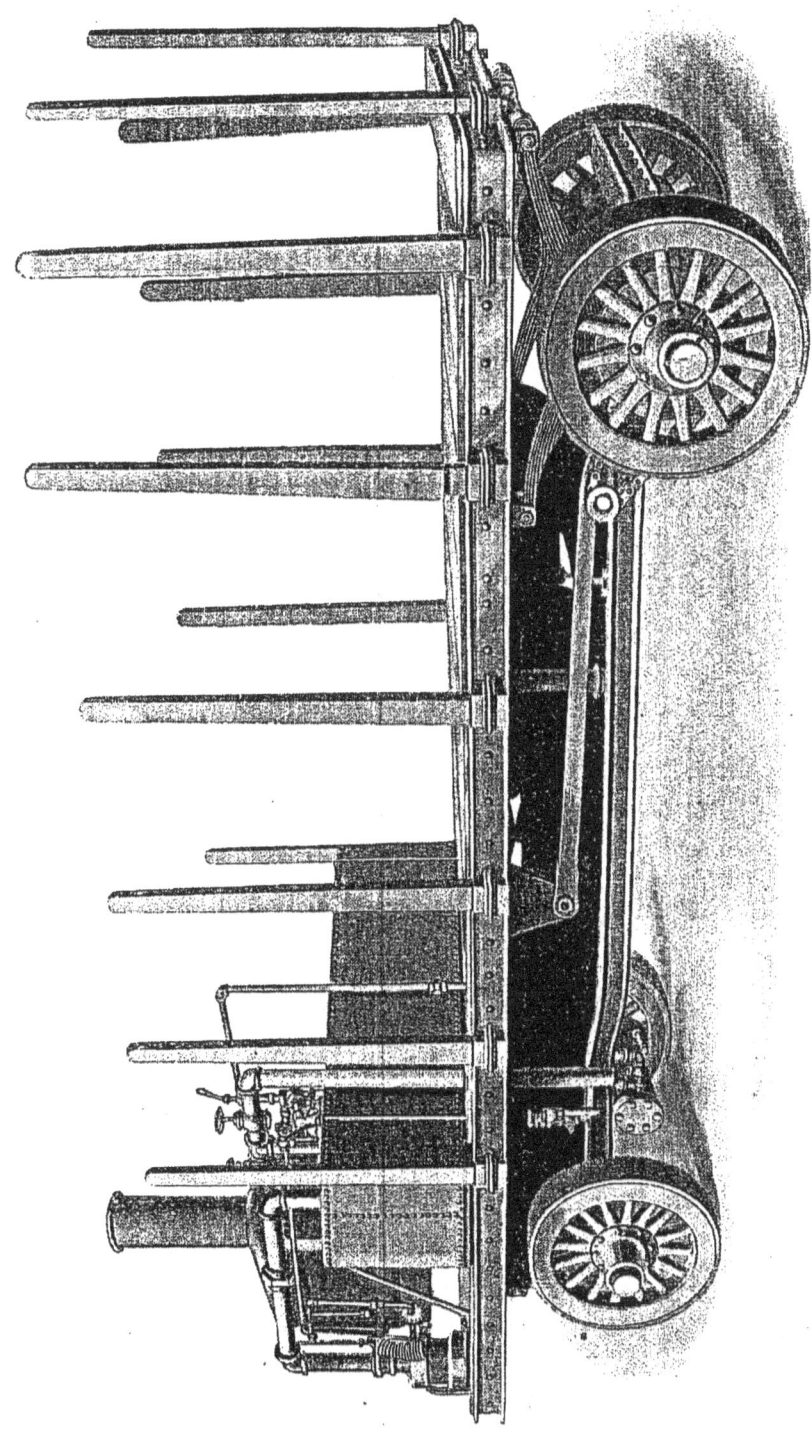

EXPERIMENTAL 4-WHEEL DRIVE STEAM WAGON BUILT BY THE AULTMAN CO., CANTON, OHIO.

74

Note the simplicity and accessibility of all parts

Do You Want a
Custom Built
STEAMER
□?

STEAM VEHICLE
DISTRIBUTORS, Inc.
GARFIELD, OHIO

WHAT WE ELIMINATE

No Spark Plugs.
No Carburetor.
No Magneto.
No Fly Wheel.
No Clutch.
No Racing of Motor.
No Gears to Shift.
No Propellor Shaft.
No Self Starter.
No Distribution.
No Vibration.
No Differential Gears.
No Complex Electric Wiring.

But we do have what we believe to be the smoothest and most powerful Vehicle for its weight in America.

Boiler, 20x16" Fire Tube.

Engine, 4 Cylinder connected directly to rear wheels.

Burner, Vaporizing Type.

Condenser, Fin and Tube.

Pumps, Balanced Type.

Throttle, Releasing Type.

Burns Kerosene or other Distolates.

Agents and Dealers wanted for Territory now available.

EXPLANATION FOR TABLE ONE.

Most of these figures are the result of painstaking research conducted by the writer among various primary and secondary resources. These numbers are intended to represent approximate output of U.S. origin only. In the absence of other data, these are the most reliable indicators available. In this vein, companies which produced only experimental models have been included; thus, one experimental steam car would be just one in the total. However, makes of any kind where no production could be documented or inferred, have been *excluded* from the table.

Of course, it must be kept in mind the figures did not emanate from a single source. A diverse group of references were utilized, which means, quite simply, no universal method of compiling numbers was possible.

A note about year-to-year car registrations by power type. A telephone conversation with Mark Patrick, then curator for the NAHC, on August 24th, 2001, was most enlightening. Mr. Patrick indicated such registration records for the early period we are primarily concerned with here are both incomplete and inconsistent. Thus, they can not be relied upon. The NAHC itself had records, at this writing, from the early period only for New England states and California. Even *these* reports do not consistently give engine type.

Finally, as far as this writer knows, this is the first attempt to document early car production by type in the United States. However, this table should be used as a general guide only.

YEAR	TOTAL	ELECTRIC	GASOLINE	STEAM
1900	4192 (100%)	1575 (38%)	936 (22%)	1681 (40%)
1901	7000 (100%)	2437 (35%)	1432 (20%)	3131 (48%)
1902	9000 (100%)	2627 (29%)	2043 (23%)	4330 (48%)
1903	11235 (100%)	2951 (26%)	6839 (61%)	1445 (13%)
1904	22142 (100%)	649 (3%)	18500 (84%)	2993 (14%)
1905	21692 (100%)	1432 (6%)	18699 (86%)	1562 (7%)
1906	33200 (100%)	1639 (5%)	30145 (91%)	1416 (4%)
1907	43000 (100%)	4027 (9%)	37002 (86%)	1971 (5%)
1908	63500 (100%)	4459 (7%)	57125 (90%)	1916 (3%)
1909	129990 (100%)	4858 (4%)	122933 (95%)	2199 (2%)
1910	181000 (100%)	7186 (4%)	171813 (95%)	2001 (1%)
1911	199139 (100%)	7400 (4%)	191218 (96%)	701 (.3%)
1912	356000 (100%)	5600 (2%)	349770 (98%)	630 (.1%)
1913	461500 (100%)	3467 (.75%)	457533 (99%)	599 (.10%)
1914	548139 (100%)	6700 (.1%)	540439 (99%)	1000 (.1%)

Totals will not equal 100% due to rounding.

EXPLANATION FOR TABLE II.
(TABLE OF EARLY AMERICAN CARS, 1805-1878.)

In this table, the writer has endeavored to create as complete—and accurate—an account of early American automotive efforts as possible. A number of sources have been consulted, although the reader is referred specifically to the monumental *The American Car Since 1775* from Automobile Quarterly and Kimes' magnificent *Standard Catalog of American Cars, 1805-1942* for further information.

Since the goal was a comprehensive listing, **all** cars for which reliable information could be gleaned are included. Also, please note only in cases where **actual** cars could be verified was the entry included.

For the historian, though, one overbearing disturbing fact remains: Who did What and When often leaves much to be desired. Nonetheless, some facts are obvious. The reader will immediately notice the decided preference for steam and electric power among such early automotive efforts. In spite of this, the internal combustion engine would early on hog the bulk of the development time/money.

Finally, without defining exactly what constitutes a car, the table is largely limited to vehicles which could run under their own power on roadways. Constructions which were obviously commercial in nature have been omitted. There have been some discrepancies with regards to dating and such. The writer has tried to relate what he believes is the most accurate of the data. We stopped the table with 1878 as this was the year of the first auto "race" in America. The car had arrived by then.

TABLE OF EARLY AMERICAN CARS, 1805-1878.

YEAR	PLACE	NAME	MOTIVE POWER	DISPOSITION OF VEHICLE
1805	Philadelphia	Orukter Oliver Evans	Steam	Not extant; sold for scrap in 1809.1
1825	Edgar County, Illinois	? Tim Parker	2 cylinder Steam	Not extant; disposition unknown.2
1826	Springfield, Massachusetts	? Tom Blanchard	Steam	Not extant; disposition unknown.3
1828	Philadelphia	Johnson (?) Johnson brothers/ Nicholas James	1 cylinder Steam	Destroyed during test; apparently.4
1828-29	New York, N.Y.	? Samuel Morey	Internal Combustion	Destroyed during the test run.5
1829	New York, N.Y.	Steam Wagon William James	Steam	Not extant; disposition unknown.6
1830	New York, N.Y.	Three Wheeler William James	Steam	Not extant; disposition unknown.7
1834	?	? Tom Davenport	Electricity; Battery Powered	Not extant; disposition unknown.8

1837 (circa)	Brattleboro, Vermont	? John Gore	2 cylinder Steam	Vehicle crashed in a ditch; abandoned.9
1847	Massachusetts	Moses Farmer	Electricity; Battery Powered	Vehicle not extant; disposition unknown.10
1851	New York, N.Y.	John K. Fisher	Steam	Not extant; disposition unknown.11
1851	?	Charles B. Page	Electricity; Battery Powered	Not extant; disposition unknown.12
1853	New York, N.Y.	Steam Wagon John Fisher	Steam	Not extant; disposition unknown.13
1857	New York, N.Y.	Dudgeon Richard Dudgeon	Steam	Destroyed in Central Palace Fire.14
1858	Hollowell, Maine Judge	Richard Rine; Miclench brothers	1 Cylinder Steam	Not extant
1863	Roxbury, Massachusetts	? Sylvester Roper	Steam	Not extant; disposition unknown.15
1866	New York, N.Y.	Dudgeon Richard Dudgeon	Steam	Vehicle still exists; in a private collection.16
1866	Bridgeport, Connecticut	? Henry House	Steam	Vehicle not extant; dismantled.17
1866	Newburyport, Massachusetts	Frank Curtis	Steam	Not extant; disposition unknown.18
1867	Bayonne, New Jersey	? Elijah Ware	Steam	Numerous vehicles; at least one exported.19
1868	Manchester, New Hampshire	James Batchelder, Will Writner	Steam	Not extant; boiler/motor sold after trial.20

1870	Bay City, Michigan	Frederick Forsythe	Springs	Not extant; disposition unknown.21
1871	Racine, Wisconsin	"The Spark" J. W. Carhart	Steam	Entirely unknown; may have been scrapped.22
1878	Oshkosh,	Oshkosh Steam F. A. Shomer, A. Gallinger, O. F. Morse	Steam	No longer extant.23
1878	Wequiock, Wisconsin	Green Bay E. P. Cowles	Steam	No longer extant.24

This has been another difficult piece of research. Part of the confusion among early sources is undoubtedly owed to the fact that, for part of the period we are concerned with, there were two rival exhibits in New York. Specifically, this was from 1906 on, when the A.L.A.M. and another entity (the A.C.A.—the Automobile Club of America) were holding sizeable displays in relatively close proximity. For the sake of continuity, we are dealing here with the Madison Square Garden show.

On a different note, there have been a number of years where different, conflicting, stats have been available. The writer has endeavored, to the best of his ability, to make as accurate a record as possible with the data that are available. When something so simple as the number of exhibitors at a nationally known event can not be stated exactly, this shows in the disarray auto history was in at this writing.

EXHIBITORS AT MADISON SQUARE GARDEN,
1900-1912.

YEAR	TOTAL	GASOLINE	STEAM	ELECTRIC
1900[1]	34	19	7	6
1901[2]	139	58	58	23
1902[3]				
1903[4]	253	168	34	51
1904	238	219	9	20
1905[5]	212	177	4	31
1906	160(?)	159	1	-
1907	61	57	1	3
1908	48	34	4	10
1909	44	32	1	11
1910	54	45	-	9
1911	60	58[6]	1	1
1912	60	56	1	3

ABBREVIATIONS/GLOSSARY OF KEY TERMS.

ACKERMAN STEERING SYSTEM: Automotive steering system, virtually universal on automobiles at this writing, that allows for turning and control of vehicle.

DOS-A-DOS SEATING: In a four passenger vehicle, the passengers sit back-to-back.

FOUR-CYCLE (BRITISH: FOUR-STROKE): An engine type which utilizes four cycles to complete combustion. It has intake and exhaust valves.

HP: Horsepower. Officially, a unit of power equal to 746 watts in U. S. standards.

IGNITION SYSTEM: The entire system that causes the vehicle to run, other than fuel.

IRONS: A hot burning iron applied, in this case, to the steamer's pilot burner.

LBS: Pounds.

L-HEAD: Engine type in which the valves are all arranged on one side of the block, so the combustion chamber is shaped like an inverted 'L.'

NAHC: The National Automotive History Collection, located in the Detroit Main Library.

NAPHTHA: Any of various volatile fuels of hydrocarbon mixtures.

PSI: Pounds per square inch.

PILOT BURNER: The pilot used on a steamer to fire/warm up the steam.

PNEUMATIC AND SOLID TIRES: A pneumatic tire uses air to cushion the ride. Solid tires were made of solid rubber, with no air pockets.

RPM'S: Revolutions per minute.

SPIRITS: An alcoholic solution of a volatile substance.

T-HEAD: An engine type in which the valves are on opposite sides of the combustion chamber in the block. Requires two camshafts. The combustion chamber resembles an inverted 'T.'

THERMAL EFFICIENCY: Concerns how quickly a system can dissipate or use heat.

TORQUE: Twisting force at the flywheel of an engine.

TWO-CYCLE (BRITISH TWO-STROKE): Engine type *without* valves; only requires two cycles to complete combustion.

V8: An engine type with eight cylinders arranged in a 'VEE,' four per side.

PERIODICALS, NEWSPAPERS, OTHER SOURCES.

PERIODICALS CONSULTED:

Antique Automobile, Automobile Quarterly, Automobile Review, Automobile Topics, Automobile Trade Journal, Business History Review, Central Station, Cleveland Plain Dealer Pictorial Magazine, Cycle & Automobile Trade Journal, Electrical World, Goggles & Gaunlets, Invention & Technology, Journal of the Franklin Institute, Journal of the Illinois State Historical Society, McClure's Magazine, Metropolitan Magazine, Motor, Motor Age, Motor World, Ohio Motor Travel, Old Timer's News, Outing Magazine, Power Wagon, Ross County Recorder, Scientific American, Theatre Magazine, The Automobile, The Automobile & Motor Review, The Horseless Age, The Motor Vehicle Review, The Ohio State Journal, The Upper Hudson Valley Automobilist.

NEWSPAPERS, ETC.:

Chillicothe Gazette, Damascus Herald, Detroit News, Rockton Herald, Scioto Gazette, The Salem News.

TABLE OF EARLY AMERICAN CARS, 1805-1878.

Internet Sources as cited in text.
(Footnotes)
1 This was the first vehicle to move under its own power in American city streets; but it weighed 20 tons. A model was in the Smithsonian Institute at this writing.
2 Three Wheeler; two rear wheels said to be 8 feet in diameter.
3 Eight seats in a wheeled carriage.
4 Probably caused America's first automobile accident.
5 America's first I. C. car; probably used Camphene for fuel.
6 Ran along NYC streets at "great speed." Speed was not specified.
7 Had drive to the rear axle.
8 First electric; used on a short track; but battery was not rechargeable.
9 Builder drove this vehicle more than a decade; had wood-burning vertical boilers.

10 Two passenger; battery had 48 Grove cells (one pint capacity), but not rechargeable.
11 Had a guaranteed speed of 15 mph; price was listed as $200.
12 Could do 19 mph, apparently; non-rechargeable battery.
13 Top speed 15 mph; vehicle in use for two years.
14 America's first *practical* steam wagon.
15 Two passenger, Four Wheeler. Two hp, 25 mph claimed. No brakes, reversed steam to stop. (Several were made).
16 Steam cylinders on each side of a horizontal boiler.
17 Vehicle weight of 1800 pounds; 4 hp, 30 mph. Three forward gears, one reverse.
18 Curtis evaded an arrest warrant by fleeing in the car (first getaway by car in American history). 5 hp, 25 mph; Price, $1000.
19 Several cars were constructed by Ware, possibly as early as 1861. One was sold to a private buyer in Canada for $300. Above example typical.
20 8 hp; one cylinder. Boiler had trouble staying lit in the car.
21 Design was patented, April 19th, 1870.
22 Car's whereabouts were known until World War Two; subsequently, it disappeared.
23 Car had 2 cylinders, 1 forward and 1 reverse speed. It won the first automobile race ever held in the U.S.
24 Car had 2 cylinders, three forward speeds, and weighed 14,255 pounds. It lost the first auto race in America against the Oshkosh Steam.

EXHIBITORS AT MADISON SQUARE GARDEN, 1900-1912.

1 Eugene Lewis, *Motor Memories* (Detroit: Alved Publishers, 1947), 19.
2 Schroeder, *World of Autos*, 25.
3 Merksamer, 14. There was no 1902 show.
4 Flink, *America Adopts*, 234-235.
5 Flammang, 40.
6 There may have been 57 gas cars and a total of two steamers at the show, instead of one like the chart shows.

The Final Analysis   The Failure of the Alternative Cars?

I. AUTOMOBILE HISTORY OFTEN SHROUDED IN UNCERTAINTY.

Now that we have examined our little-known alternative car makes, we need to briefly pass over the bigger picture of just why all of this happened, back there at the start of the twentieth century. Some essential questions need considered at this point. Inquiries like, 'Was there some basic flaw or two that rendered the alternative cars into an undesirable motive source?' 'Did they possess erstwhile hidden shortcomings that doomed them from the start?' 'Or, did they indeed receive the proverbial 'raw deal' as has often been alleged?'

We have a major handicap here, though, as much of the information we could have used is no longer accessible in any form. It is possible some undiscovered treasure trove may yet come to light. But, at this writing, there were wide gaps between what we know for sure and what was largely conjecture about those early far-off days.

An obvious way around this problem, i.e. interviewing some of these pioneers, presents a formidable obstacle of its own: They are all dead! This *really* limits our available primary resources. It follows, then, a number of glaring omissions and errors have inevitably crept into the record, impeding genuine historical research in the process. In fact, "Automotive history is full of errors and inaccuracies which will never be fully corrected."[219] In short, the average historiographer, at this writing, tended to question more and accept less at face value as a result. There was little choice in the matter.

Compounding the problem, few of those early "automobilists"[220] were exceptional record keepers.[221] They were likely too busy making history to see the need of recording it. For instance, Frank B. Stearns (1879-1955), builder of the Stearns make of cars from Cleveland, seems fairly typical of that hardy early breed. He minced no words.

"Between 1890 and 1900 we turned out in our old barn about fifty cars, [and] we kept no books but a checkbook and the office was in my hat."[222] If that were not bad enough, we have to be careful about using those few pioneer accounts we *do* have. The natural human tendency to overstate one's own accomplishments at the expense of others comes into play here.

Also, most of the available accounts by the pioneers themselves were first put to paper years after the fact, when memories were dimming and aging garrulousness was impending. Other pioneers who could have written memoirs lacked either the time, the enthusiasm, or the education to actually do so. However, even with all the flaws, there can be no substitute for these valuable primary resources. It should be immediately obvious how difficult it was to extract the pertinent historical facts from such a sketchy record.

Two things are certain and essential to this argument: 1.) The passage of time has dimmed the memory of the alternative cars. As a society, we may have even forgotten there was once another choice or two other than the gasoline engine for an automotive power source; 2.) Having stated that, it is equally true the internal combustion engine early on established itself as the consumer's favorite. There is no getting around either very basic point.

But should we not make the mistake of thinking the triumph of gas cars even *then* was a foregone conclusion. It must be borne in mind, in none of the three main power sources was the full extent of the newly emerging technology available at once. In fact, "technical changes. . . came slowly and patchily, not tidily in one great enlightenment."[223] Thus, no one knew which type—if any particular one—would beat out the others, until they had all reached some degree of mechanical "perfection." This would inevitably take time to sort out.

Nor was this a simple matter. As a contemporary writer stated: "Each [type] has its peculiar advantages and spheres of use, and the first thing. . . is to recognize frankly its limitations, and to set about perfecting it within those bounds."[224] And even at that early date there was no shortage of firms engaging themselves in the manufacture of cars. "At least eighty establishments are now [in 1899] actually engaged in [auto manufacture]. . . representing no fewer than two hundred different types of vehicles, with nearly half as many methods of propulsion."[225] Baker may have been exaggerating for effect; but the one constant remains the gasoline engine was just, at that point, a single form of motive power available among many venues.[226]

II. *WERE THERE OTHERS?*

We are not saying there were no other realistic alternative power sources. For instance, a company called American Pneumatic was incorporated in West Virginia near the end of the nineteenth century, and, in February 1900, announced it was going to build cars powered by compressed air. Instead, the company was apparently made *of* air; there is no evidence of even a single car.[227] By the way, cars using compressed air at this writing had the same problem they had back at the beginning: high costs associated with the fuel and stations to receive it at.[228]

Although the three main types definitely were more common, there remained a few different power sources that would never be mainstream. Moreover, there were even cars "powered" by springs, at least in theory. In 1895, an Andrews' spring car of Center Point, Iowa, was apparently built. It turned out to be a baby carriage, pure and simple. The "engine" was said to produce six horsepower.[229] Now, to be fair, this spring construction did provide some success for relatively simple constructions like a baby carriage. When the designer, one A. B. Andrews, tried to apply his spring technology to cars, however, he utterly failed. The obvious difficulty in this case has to do with the laws of physics; i.e., the bulk and weight of even the earliest cars was so much greater than a baby carriage. But Andrews was stubborn. He announced his intention to enter his vehicle in the *Times Herald* race of 1895, "but was unable to raise the necessary money" to perfect his spring motor.[230] It was just as well, for the impracticality is obvious. Spring-powered cars were really in vogue only for a fleeting moment anyhow. Still, "more than fifty attempts were made to build them."[231]

There were also more personal forms of transportation. In 1906, one Alphonse Constantini invented motorized roller skates. He felt the need to make a contribution to modern motoring. Or, at least, for people interested in more personal forms of transportation. Apparently a person with the skates could do up to 40 mph. Despite this well-intentioned designer, though, this particular device lacked staying power.[232] Its fatal nemesis appears to have been the annoying tendency for one of the motors to stall, leaving the wearer with just one spinning skate motor.

Another handy dandy device was the rotary engine. "Mounted directly on each wheel so that no gears, drive shafts, or clutches were needed."[233] The problem here was, 'how to design a universal motor that would fit a wide variety of vehicles?' Obviously there was no way to make a single apparatus work for all kinds of motor carriages. In practical terms, there were no easy answers to the problem.

Actually none of the spring motors, rotary engines, or motorized skates were ever practical. Wind propelled road carriages, and others of that ilk to the side, it becomes apparent that a great number of different ideas were tried at the dawn of the automobile age. No one type stood apart from the others at that stage. And it was not for a lack of trying. Obvious mechanical skill on the part of the operator was another factor in these alternatives.

Continuing with this line of thought, we will stretch the term 'automobile' out of shape to consider one of the more unusual contraptions in the realm of transportation. The following will show often the power sources were not confined to wheeled transportation. In 1870, in Newark, New Jersey, two gentlemen secured a patent for their invention. Zadoc P. Dederick and Isaac Grass, two quite enterprising locals, conjured up a steam model of a full-size man, complete with top hat. It was powered by a two-cylinder steam engine, fired from a boiler behind the jacket, and utilized a series of cranks to work the complicated legs.[234] A two-speed transmission, featuring also two speeds in reverse, was fitted. Apparently, this 'Steam Man' was designed to pull a cart/carriage behind itself. Normally, we might consign such a yarn to fiction, but no less an authority than Charles Edgar Duryea (1862-1938),[235] though, stated he saw the invention in action.[236] The disposition of the 'Steam Man' was unknown at this writing.

III.   *OTHER CONSIDERATIONS.*

Also, we must not overlook the fact that motive power, whatever form it would ultimately take, was just one factor that needed worked out before cars became more mainstream. There were other issues like, 'how many wheels?,' choice of gearing, braking, etc. All of these, and more, were of real concern. As one source offered, "inventors had to equally devote their attention to these matters."[237]

We must state here, though, we are concerned with just the facts. Too often, looking back to the 'Good Ole Days' wraps them in a misleading, nostalgic aura that subsumes facts and lore into an intricately wound coil. The distinct boundaries between the two become blurred in the process. Thus, emerges the danger of 'what if,' or counterfactual history.[238]

Nevertheless, there is an old myth we need to dispel while we are at it. The allegation has been made the alternative cars, although *proven* beyond doubt to be better as power sources than the gas motor, were driven into extinction by a conspiracy dreamed up between the government and the business interests in Detroit. But, to put it plainly, "The steamer was not (as some steam nuts [sic] darkly hint) put out of business by plots devised by Detroit."[239] For that matter, neither was the electric.

Why can we be so certain? Such a conspiracy was not needed. The alternative cars had enough disadvantages by themselves to see to their own demise.[240] There was thus little incentive for such a diabolical conspiracy on behalf of 'big oil.'[241] Besides all of this, the field lacked key players yet. The idea of the Detroit colossuses was still off in the future. In 1908, for instance, when General Motors was founded, the alternative cars already were just 10% of the new car market. And they were fading away fast, we might add.

Another mitigating factor, this one often overlooked by people for whom e-mail and instantaneous communication have been taken for granted, is often cited, although, once again, with little foundation. Morris McNeal Musselman (1899-1952), whose own generation was among the last for whom the automobile was still new, said it best. "They [speaking of the early auto pioneers] had little to guide them except hearsay and they solved their problems with no encouragement from the world around them."[242] At first glance, this might explain some of the wide discrepancies in the automobile record.

IV. *ALTERNATIVE CARS BEATEN BEFORE SELF-STARTER'S ARRIVAL.*

We have already mentioned an underlying theme throughout several of the consulted sources to the affect the alternative cars were adversely rent by the arrival of the self-starter in 1912.[243] The writer contends this was not the case at all, this being the first of his major arguments.[244]   There have been formidable obstacles along the way to this conclusion.

Naturally, numbers could tell part of the story here. Unfortunately, production figures for most of the model years we are concerned with now are rather scanty, although the writer has spent considerable time trying to reconstruct them, culling the data from a wide variety of sources in the process.[245]

We have two constants here as well: 1.) In the beginning, the steamers and electrics made up a substantial part of the new car market. That much we have already established. As such, there are quite a number of "firsts" in the car world that are associated with the alternative cars. On May 20th, 1899, an operator of an electric hansom cab, one Jacob German, "enjoyed" the rather dubious distinction of being the first person ever arrested in world history for violating a posted speed limit. That of New York City; which had an eight mph limit. German was doing a rollicking 12 mph at the time.[246]   A few months later, on September 13th, 1899, in the first recorded auto fatality in this country, an electric car of unspecified make driven by one Arthur Smith struck and killed a pedestrian named Henry B. Bliss on the crowded streets of New York City.[247] Then, 2.) The market penetration for the alternative cars deteriorated both quickly and almost completely. By 1912, the year the Kettering Self-Starter first came to market, only one steam car company, the Stanley firm, was still building steam cars in any appreciable numbers. Even at that, *they* made a mere 566 cars that year.[248] Gas car production simultaneously was a whopping 98.25% of the total United States automobile output of 356,000 (349,770).[249] Comparisons here are very telling. New electric registrations in 1912 were a mere 5,550.[250]

There was a more ominous, underlying problem. With just one choice of "popular" make in 1912, the steamers had little variety of selection. The lackluster sales of this type can thus be partially explained away. But, 'why were the electric cars not selling?' It certainly was **not** because of a lack of different models. In fact, "Ninety-four electric models were offered by twenty different companies."[251] In spite of this, a check of all possible sources betray total production could not have exceeded 5600 cars. Or, approximately 1.6% of a vast U. S. output.

Obviously by that point, then, neither alternative type was in a position to unseat the sheer dominance of the gas car, barring a latter-day miracle.

It must be borne in mind the practical self-starter did not even appear **until** that year. With that, it becomes clear the gas car had long before this started to unseat the alternative cars. Equally obvious there were distinct disadvantages between and among the alternative cars, independent of the absence of a self-starter on the gas cars.[252] Unrealistic expectations had their place in this whole affair.

As one reporter succinctly put the matter: "Give us an electric passenger automobile that will travel 100 to 150 miles at a reasonable speed on a single charge of the battery. . . and hundreds will change from gasoline to the electric vehicle."[253] Others praised the steamer's ease of operation once the car was ready to roll, but denounced the heavy prepping involved beforehand. But, we need not get ahead of ourselves. For we must observe *each* of the three main types had its advantages and disadvantages. Some inventors, for example, loved the gasoline engine.

One of these men, Hiram Percy Maxim, wrote: "The thing which interests me today [in 1936], after nearly forty years have rolled around,. . . [is we] insisted upon addressing ourselves to the development of the gasoline engine rather than making use of the already highly developed steam-engine and steam-boiler."[254] Apparently, he was interested solely in gasoline engines. "The only form of motive power that would suit me was some kind of an engine which would. . . [go] as long as its fuel tanks had anything in them."[255]

Ironically, Maxim worked for the same George Pope who would later build the electric-driven Pope cars.[256] "Neither Albert A. Pope nor George Day [general manager] saw any future for the gasoline car and preferred to concentrate on electrics."[257]

Some pioneer motorists did not mind the internal combustion cars. There were other more particular individuals; men who loathed the 'gassers.' They went on record blasting the gas car as, 'noisy, smelly, moody, greasy, and unreliable.' Or, just plain 'uncivilized.' This was all true, at the outset; and the much-maligned alternatives had *some* attractions.[258]

Proponents of steamers proudly pointed to one clear-cut advantage. The steam engine, since its power source is not internally driven, can be readily useful at almost zero rpms. At low rpms, even the earliest steamers were capable of producing sufficient power to shove in a wall. Technically, the steamer's engine does not even produce power; since it has external combustion, it merely translates the available power made by the boiler into motion. This released power is then available on both ends of the piston, so steam engines can be very smooth running machines. No uneven power curves with all the attendant problems.

It follows the steamer can be developing power without turning, once fired up, of course. In sharp contrast, the gasoline engine is internally fired. It *must* be turning to be developing power, besides spinning at a healthy rpm. That is because the power is produced by explosions within the cylinders, not energized in the boiler and *then* put into them.

In the beginning, these differences were a serious shortcoming for the gas cars. In addition, the internal combustion car had to be turning fast to even start. This was difficult to achieve by hand.

V.   THE 'AGE OF THE GIANTS.'

In fact, the internal combustion engines have always needed an outside source to fire them up, which is why a working self-starter had become of such importance in the first place. The writer contends improvements for the gasoline engine came quickly, but not in this one particular area. Still, from about 1904, filling this void became a very important work in progress. Almost from the start, desires had been expressed to develop a such a device. Then, the calls became **necessary.** And we must keep in mind convenience was of secondary importance here.

Looking at a good Cleveland, Ohio, marque like the Peerless Motor Car Company reinforces this notion. In 1904, that make's biggest engine was a four cylinder T-Head of 227 cubic inches (4.25

in. bore X 4.00 in. stroke). Not a small engine, but not overly large either. But for the 1912 model year, just eight years after, the largest Peerless engine had swelled to a massive six cylinder T-Head of 825 cubic inches (5 in. bore X 7 in. stroke).

Why the increase? Peerless was just following an industry practice among luxury car makers. Whatever the reason, though, this became, in short order, a serious obstacle. In 1904, trying to hand crank the medium sized four cylinder must have been daunting, so how could one handle the 825 six? The latter would likely be above the ability of most men and practically all women to achieve.[259]

It is no coincidence the very next year, company models were using a powerful compressed air starter. A strong device indeed would be needed to spin that six-cylinder monster. In fact, "a [1913 model] Peerless weighing 5100 pounds was driven 660 feet up a 7 percent grade in 12 minutes on the starter alone, without exhausting the battery."[260]

This situation can be blamed at least in part on consumers, who demanded bigger chassis, bodies with more elaborate trimmings, and greater girth almost with each passing year. This, in turn, led engineers to boost engine size, often substantially as we have seen, to maintain a comfortable level of efficiency. Then, in a quest for increased road performance, engine dimensions grew again as a general rule, especially among the high-priced market. We have already observed the gasoline engine was lagging behind its alternative cousins in terms of technology in the early period. To compensate for this lag, engineers had perforce to construct engines of gargantuan dimensions.

All of these factors thus contributed to the jumbo size of some of these early gas engines. The Peerless engine cited above was as large as a production gasoline engine would ever get, though. Pierce-Arrow, another fine car manufacturer, shared the same size production engine with its biggest luxury models. We should note here, however, most of the early gas engines were about one-third smaller in displacement than this at their peak on the average.

This is the second of the writer's three tenets: That the gas cars, although initially inferior in technology to their alternative cousins, quickly made up for it. Said James Melton (1902-1961), noted opera singer and early antique car collector, the gas engine is, "very badly suited to the task of running an automobile."[261] But this type in its basic form also readily responded to improvements. Even Melton, who was a steam car maniac, freely admitted that much.[262]

Nonetheless, this so-called "Age of the Giants," which began around 1904 and lasted until about 1915-1916, really provided the impetus at that precise moment in history for the development of a practical self-starter. Very important from an historian's perspective, for automobiles would never be viewed quite the same way again.

In contrast to all of this, the steam engine, except for a few notable detail improvements in condenser technology and the like, was much more impressive at the beginning.

To reiterate, an early gas motor had to be of immense size to provide the power output of a very small steam engine. Further, the steam engine was much more compact than contemporary gasoline engines. On the surface, in fact, the steam engine's advantages may have appeared overwhelming. However, the latter's ultimate potential—and appeal—was to be really limited by comparison with the gas engine.[263]

As for the electric motors, they played by a different set of rules: their power tended to be constricted by the size of their battery's cells and ratings. In their case, weight rather than a complicated start-up like the steamer's nemesis was the premiere detrimental factor.[264]

VI. *A 'WEIGHTY' ISSUE.*

The ponderous bulk of the earliest electrics really led to other problems. For instance, tires—often overlooked from a more modern perspective—were a real headache for early electric car concerns. In early tests, those narrow tires and wooden spoke wheels of the so-called first generation electric cabs had proven adequate. That is, manageable in Philadelphia's main streets where the pioneering horseless carriages were practically developed as a business venture by Pedro G. Salom (1856-1945) and Henry G. Morris (1840-1915). Deficiencies showed up quickly, however, on less friendly surfaces. There was, in this latter more rugged venue, a singularly high rate of tire fatigue/failure.

An unavoidable consequence. The electric cars had to venture beyond Main Street, U. S. A. Operators quite methodically were trying to displace the older horse-driven hansom cabs from their dominant position in the lucrative realm of public/private transportation.

Nonetheless, when these same tires/wheels proved thus unsatisfactory (i.e., on those less than perfect roads outside of big cities like Philadelphia), newer designs had to be substituted. There were a couple of viable reasons why. The big one was 'Weight.'

We have established electric cars and their powerplants were heavy. So were steamers. Surprisingly, steam engines themselves were relatively light. For example, the Stanley's engine weighed a mere 205 pounds by itself.[265] If that *were* the only factor in the weight equation, this would be an impressive achievement indeed. But the necessary accompanying tubing, in conjunction with the boiler, really piled on the weight, in the process negating some of that potential weight "saving."

If weight was of moderate concern for steamers, though, it was a *real* problem for electrics. A host of the troubles with the electric cars' tires, as we have hinted, were due to those heavy, bulky batteries. The latter remained heavy because battery technology was still in its infancy at that early developmental stage.

Under the strain of this difficult work of powering cars, the batteries perforce had a tendency to, "fume, give out on the road, leak. . . disintegrate, corrode, short-circuit."[266] This does far more than suggest the early battery types fully deserved their bad reputation. As a form of remedy, later electrics offered redesigned battery compartments, among other perks, in order to facilitate removal of fumes from this critical area.

It is significant only solid rubber tires—graced with the misnomer of 'cushion tires'—were at first employed on the ponderous electrics.[267] Later, however, "tire manufacturers were able to develop a pneumatic tire capable of withstanding the jolting weight of electrics."[268]

In contrast, some of the earliest steamers were really lightweight. The new Locomobile, for instance, was a then revolutionary type of buggy from around the turn of the century. The type was lightweight and rather simple to put together. It had been road tested on the fine thoroughfares around Boston, where the cars actually thrived for a brief period. But, the slight contraptions, "quickly shook to pieces on the miserable trails, called roads, in most other parts of the country."[269] And, this was not *just* "simple" tire failure here.

Nevertheless, it did not take long for open-minded businessmen[270] to see the intrinsic value of the new technology. As we have mentioned, Salom and Morris put out that first fleet of electric cars, in the guise of cabs.[271] These were Columbia electrics, complete with batteries that seemed to weigh almost a ton by themselves. Very quickly, though, these particular cars were challenged by more capable gasoline rivals. What was the result? "The faster, lighter, cheaper gasoline taxi soon *drove them off the streets.*"[italics mine][272]

Moreover, because of the enormous weight factor inherent in the electric car design, "The cost of repairs on tires in cab service often times exceed[ed] that of any other operating expense."[273] As for the batteries, many firms (both inside and outside of the auto industry) were hard at work experimenting and testing, trying to improve the future state of the art. Not all of this was futile, to be sure, but contemporary state of the art was certainly less than desirable.

The underlying problems persisted. How many potential sales of new electric vehicles were lost permanently waiting for the oft promised "improvements" will never be known. Moreover, this situation would never substantially change. As one editor so sagely observed, "No modern invention has enlisted so large an expenditure of time and money with so little result as the electric storage battery."[274] The solution seemed always just ahead, just around the next corner.

The charging stations were another imposing obstacle.[275] Such public facilities as existed were invariably urban, and confined to rather large metropolitan areas at that.[276] The inevitable cost factor must also be addressed with this regard to recharging. "These [charging stations] were accessible twenty-four hours a day every day of the year, and the rate was 10 cents per kilowatt."[277]

There was yet another not insignificant difficulty. "A totally depleted battery required approximately 33 kilowatt-hours of alternating current to fully recharge."[278] This situation would thus require well over a full day to recharge; meanwhile the would-be operator was on his own for that time. As Richard Wager (1939-) put the matter regarding this, one of the electric car's biggest drawbacks, "Quick-charging devices did not come until decades later, and the dreamed-of network of corner stations [i.e., for servicing]. . . batteries failed to [ever] materialize."[279] How all of this was received when the neighbor down the road merely had to fire up his gasoline car, even if by hand cranking until 1912, and he was on merry way can be imagined.

This was an especially odious result for the electric cabs. While the gasoline car quickly demonstrated its superiority for most general uses, its electric cousin was able to maintain a brief hold on the cab/delivery market. The electrics, initially at least, proved very lucrative in plying their trade. The pages of *Horseless Age* reported on the first eighteen weeks of the Morris and Salom vehicle operations. Total distance covered by the cabs was reported as 14,459 miles—an average of more than 1200 miles per cab. Total passengers claimed was 4765.[280] Interestingly, the EVC chose not to reveal the extent of the profits they derived from all of this effort; but the return was probably minimal at best.

A tangible question remains, 'Were the electric cabs popular?' Fred Vieweg, general manager of the EVC, thought they were very well-liked. So much so, in fact, one could not be procured for an evening, "later than ten o'clock in the morning."[281] Even when the national company collapsed soon thereafter, the New York branch continued to prosper.[282] When, in 1904, a new manager named Richard W. Meade arrived, he took charge of The New York Transportation Company. This firm, formerly the local branch of the EVC, was largely responsible for public transportation. Most of the cabs they were using about then—December of 1901—were electrics. In a moment we will look at an overview of the whole firm.

In urban areas, those people who owned electric cars quickly realized their limitations. They simply learned to work within the restrictions. For instance, keeping the batteries charged was a task in and of itself. This was at the head of the list. Besides this, those expensive batteries often required replacement after just three years of service.[283] Additionally, since electric motors do not produce torque in the traditional sense, hill-climbing could be a real chore for electrics.

Finally, changing the heavy batteries generally required a crane or a hoist assembly of some kind. We might mention, these batteries were not the same as car batteries of a century later.[284]

All of this leads to another interesting query. 'Was the electric car ever considered the whole answer to the motive needs of consumers?' It apparently was not, even in the beginning. Electrics just had too many limitations for that. One journal writer wrote of a notable weakness in its January 1900 issue: "There is no question of the reliability of electricity. . . but frequent renewals of the supply makes it inferior as a means of locomotion for long distances." Significantly, this far-seeing writer concluded, well over a decade before the self-starter appeared, although the electrics were then the most popular, "the automobile tourist—for he will be a fixture in America as he is abroad—will use a steam or gasoline motor for long runs."[285]   The prophecy was indubitably right. As it would turn out, those who put their faith in the electric car as their prime mode of transportation were doomed to both early and all too frequent disappointment.

In fact, prolific inventor Thomas Alva Edison (1847-1931) freely made known his opinions about which type of power source would eventually triumph. When he was queried in 1895 about the future of the electric car, the electric guru remarkably—and prophetically—stated, "[I]t would seem more likely that they [speaking of future cars] will be run by a gasoline or naphtha motor of some kind."[286]

The above did not prevent Mr. Edison from appearing frequently in public in an electric car. His motives were generally less than altruistic, though. "Edison lent his name. . . not entirely because he had confidence in their [electric cars'] future but because most of them used batteries made by his company."[287]

Not that the inventor failed to try his best to perfect a battery. On May 28th, 1902, Edison's firm announced a new type of battery to keep electric cars competitive. He felt reasonably certain this new nickel-alkali battery would, at the worst, keep electric cars competitive. At best, it might just make possible the triumph of this one type over the others. In the end, it did neither.

In the end, the besieged Edison helped dig the grave for the electric car, without ever meaning to. When he cast doubt about the ability of the electric car to carry out its duties, just as the internal combustion car was starting to jell, the die was cast. After all, who would know this matter better than an electrical expert?

However, since most electrics were used for shopping by urban ladies, and their speeds at the end were approaching 30 mph—just enough to keep up with the ever thickening flow of city traffic—the shortcomings were at least liveable, for a time.[288] Thus, within their narrow confines, the electrics briefly thrived as residential cars. But consumers at large just never accepted them.

VII. *THE ELECTRIC VEHICLE COMPANY.*

How about the commercial opportunities for electric vehicles, though? Those long-range prospects were not much better. For instance, the Electric Vehicle Company, formed by a combine between our old friend Colonel Albert Pope and William C. Whitney (1841-1904), was doomed to be a victim of its own success. A New York City financier and a former Secretary of the Navy under Grover

Cleveland, Whitney was busy by the mid-1890's in the process of cornering the public utilities market of America's largest city. The industrialist envisioned fleets of company-owned electric cars serving as cabs and commercial buses catering to an eager American public. Whitney was singularly ambitious; he planned eventually to have such fleets in every major American city, taking full advantage of this new form of transportation.[289] After all, with the automobile still in its infancy, there were few ground rules yet about how to employ the new conveyances. Many possible courses remained open to those gifted with ambition.

So Whitney's plans were grandiose and included a variety of transportation guises, including streetcars, omnibuses, and individual electric cabs. All of these would utilize electric power provided, of course, by his own electric company. No wonder critics began decrying the combine as a 'monopoly' so soon as it appeared. But, alas, there was to be no permanent success. "By the end of 1901[,] the organization was in disarray... and the larger automobile market had swung decisively towards private ownership, long-distance touring, and internal combustion."[290]

What was the problem? Well, for one thing, styling was a stumbling block. Pictures of those first cabs demonstrate their resemblance to the horse-drawn variety, and they were visible on streets where at least some horse-drawn conveyances were still around. In fact, one contemporary street scene, set in New York in 1909, had a race between the equine cab and its mechanical brethren and noted forlornly, "passersby [sic] are not aware that soon all this will be another lost moment in history."[291]   For another, there were very few automobile manufacturers around. When the "monopoly" got underway in earnest in March 1897, there were just three car manufacturers reporting business all across the United States, according to one source. The production of all three combined did not reach double figures.[292]

In order to make an effort at their far-flung plan, Pope and Whitney thus *had* to order from practically every available source. The results were all too predictable. With no standard electric cab models or even a standard battery, the resulting multiplicity of parts and services must have been daunting. This was a specific concern of company servicemen and service managers, but it was also a demonstration of more deeply rooted problems. Not until 1900, after more than 2 ½ crucial years had elapsed, was the desired uniformity of components achieved with any appreciable degree of success.[293] By then, it was far too late!

This flies in the face of the cold, impersonal logic of the Industrial Revolution, with its central idea of the interchangeability of parts.

Furthermore, the Pope-Whitney combine got into some complicated legal issues over their 'monopoly.'[294] The court battles that resulted among various and sundry concerns is not of interest for us.[295] A brief summary of the EVC *is* of concern here, though, for its method of operation could easily have become the accepted norm.

Unfortunately, in brief, the EVC experienced an increasing hardship in doing their electric car business. This failure happened for reasons which are still not quite clear. Kirsch concludes, "despite its extensive network of suppliers, employees, and consumers, the whole was less than the sum of its parts."[296] Whatever the cause, the end results were almost predictable. Shortly after, a trade periodical carried the following gloomy announcement: "E[lectric] V[ehicle] Company Placed in Receiver's Hands."[297]

Then, in what must be one of the biggest ironies of the whole situation, the EVC, its principal business of promoting electric cars a failure, turned to more lucrative financial opportunities. It pulled the Selden Patent out of its repertoire collection, and survived for a while by collecting royalty payments from the licensed gasoline car manufacturers.[298] It thus capitalized on the success of the gas car, which enabled the firm to take indirect advantage of the very motive power source that had doomed its own electrics to virtual extinction in the first place. Talk about trying to cover all the bases. Finally, what about those 'lost opportunities?'

The New York City branch of the EVC survived for a time after the parent company bit the dust. This provides a perfect venue to see how the later electric cabs actually fared in service. In actual practice, then, they received short shrift. It became more difficult to secure both new cabs and replacement electric parts for existing fleet vehicles as the first decade wore on.[299] The results were inevitable. It was not long before this facet of the firm closed up as well. Gas engines, even without a popular self-starter, were proving to be very tough, resilient competitors. And, they were about to get one. There sure is a tale behind that one.

VIII. *GASOLINE ENGINES GET A SELF-STARTER.*

If we are to contend the self-starter's contribution was negligible, we must first relate the pertinent "facts" about the invention itself. In the winter of 1910, a Cadillac stalled on the old wooden bridge at Belle-Isle located at the Detroit river. The hapless, conveniently unidentified, woman motorist was unable to get the car started again, as she lacked the strength to work the hand crank.

In a chivalrous gesture, a passing industrialist, a friend of Cadillac's Henry Leland, tried to crank the car. The gentleman was none other than Byron T. Carter—for whom the Cartercar was named. The spark had not been retarded, and the, "crank kicked back. . . striking him in the jaw with such force that he ultimately died."[300] When Leland learned of his friend's death, he exclaimed: "The Cadillac car will kill no more men if we can help it."[301] Cadillac soon had the first modern self-starter as a direct result of this unfortunate event.

This quaint tale would be a woeful one, indeed; if it were a real event. In fact, it is just *another* of those many misconceptions we have already mentioned. Errors that have started and then often taken on a life of their own from there. Even *The Encyclopedia of American Business History*, as solid a reference source as it is, has fallen victim to the myth. At one point, it states, "Carter did die of pneumonia and. . . if there was an accident involving a crank that made Leland encourage Kettering it was. . . not [Byron] Carter, who was the victim."[302] However, in the same reference, Northcoate Hamilton relates: "In the summer of 1910. . . Byron Carter was injured when he stopped to help a woman restart her stalled engine."[303]

So which version is correct? Did Byron Carter play an important role in developing the Self-Starter? Or not? *The World Guide to Automobile Manufacturers* helps solve the controversy. It states:

"According to folklore, Byron T. Carter died after being struck by a recoiling starting handle. . . [while he] actually died of pneumonia, in April 1908."[304] This was long before the businessman supposedly "suffered" his mortal injury on the bridge. Obviously such stories play fast and loose with the facts.

There are some related pertinent facts that are beyond dispute, however. For one, Charles Kettering (1876-1958), a General Motors engineer as well as a graduate of The Ohio State University, developed his starting device by patterning it after a small electric motor he had first perfected for cash registers. Also, Kettering's was not the first automotive self-starter, by any means.[305] But it was the most perfected one available. The industry would later shamelessly "borrow" Kettering's model for its working pattern, generally without giving due credit.

Second, as the gasoline engine started to gain favor, demand for its fuel took off. In a remarkably short time, gas went from a substance which had little use/demand to one of the hottest commodities around. In the beginning, almost no one even carried gasoline, and there were no service stations.[306] It can not be said with confidence even all hardware stores were actively carrying supplies of gasoline. Percy Maxim, for one, stated it was a real adventure to acquire the gasoline he needed. "I visited a paint shop. . . and asked that an eight ounce bottle be filled. The proprietor looked me over as though purchasing gasoline were a highly suspicious proceeding."[307] What if Maxim had been female, in addition to being 'strange?'

IX. *'THE FAIRER SEX' AND 'THEIR' CARS.*

In 1900, it was a man's world. Very few women drove, or, for that matter, worked outside of the home. In most states, the 'fairer sex' did not yet have the right to vote. We might briefly explore how the rise of the car and the rise of women as motorists more or less coincided. Some of the early trailblazers were *really* pioneers in more ways than one.

For instance, one can not help but admire the courage of Mrs. Newton J. Cuneo. This rugged individual not only dared drive on the roads, she participated in male-dominated road racing. She participated in the first Glidden Tour, in July, 1905. The participants drove out of New York City. Mrs. Cuneo was driving a 15 hp White Steamer.[308] Just outside of Greenwich, Connecticut, she apparently, "drove right through the rails of a bridge." [309]   Once righted, and nothing daunted, the intrepid lady and her steamer, "drove out of the [shallow] water under her own power and continued the run."[310] Drivers like this proved that to be good did not necessarily mean being male, much to the chagrin of most of her competition. In 1909, during a grueling three-day race meet at the New Orleans Mardi Gras celebration, Mrs. Cuneo bested a whole field of male competitors. She came in second, "with only Ralph DePalma (1883-1956) ahead of her."[311] But she was an uncommon female participant.

The 1905 Glidden Tour will be forever memorable as the first known instance on record of the appearance of that bane of drivers ever since: the Speed Trap! This particular one was at Leicester, Massachusetts. A particularly steep hill in the northern part of town required those early car operators to speed up enough to, "get a run at it." The local speed limit was 20 mph, actually quite generous for the time. Nevertheless, this was just not fast enough for low powered vehicles. When the Glidden participants came through, the local constabulary were waiting. "Six drivers were arrested and posted bond."[312] If one is going to do something wrong, at least do it right! Three of the six tourists, including Mrs. Cuneo, happened to drive White Steamers. After the fines were settled up, the drivers made a rather noisy protest with their cars as they left town.

Yes, Mrs. Cuneo was one stalwart individual. Lady motorists were so rare, in fact, that *Outing Magazine,* in its April, 1910, issue, ran a story about Joan Cuneo and other early women motorists.[313]

In fact, she had owned the White Steamer for just a week when she entered it in the Glidden. Alas, the times were just not ready for pioneer women racers.

After Mrs. Cuneo's stellar performance at New Orleans, there were changes. She had beaten great racers like George Robertson (1884-1955), Louis Strang, "Wild Bob" William Burman (1884-1916), among others. These men were masters in their field. The response from the powers that be was both swift and unpleasant.

The American Automobile Association promptly, "adopted a rule that no woman should in future be allowed to drive, or even ride, in a car in any of their contests."[314] Mrs. Cuneo, who took all of this in stride, commented, 'would that I could cultivate some suffragette tendencies. . . but I drive and race just for the love of it all."[315] As an afterthought, we can report the daring pioneer owned no cars but steamers until the spring of 1907. Then she purchased a gas car. This was not a racer, though. And the electric never seems to have appealed to her.

Still, women preferred electrics. Probably for reasons touched on by Robert Karolevitz. The fashions of the day often dictated, "multiple petticoats and floor-sweeping dresses," which made working the feet practically impossible through, "a dozen ells of underclothing."[316] Cars even then generally used pedals on the floor to stop and go, so this was a problem. Of course, the electric was closest to the horse in form, so it gave the illusion of having been around even longer than it had. The transition between the buggy carriage and the electric car was not a wide one.

X. *LIVING WITH ELECTRICS AND STEAMERS.*

The fact is, the electrics had a long head start. They had been there before. The earliest of these cars really stand out. They were often so well equipped with creature comforts they resembled boudoirs, which was no accident. Electric car manufacturers were quick to realize women would be most likely to occupy their car while in use. In response, these automakers did their collective bests to please members of the fairer sex as their needs were then understood in a far more innocuous time period.

The cozy ambient atmosphere of a courting parlor was almost inevitable under the circumstances. This living room environment inevitably made for a tall machine. It could be most often seen negotiating crowded city streets, with the curtains drawn, while the occupants literally enjoyed the comforts of home. In fact, "most electrics had boxy, carriage-like appearance which changed little during the 20 years they were popular."[317]

Body styling that was at best odd-ball is a common thread of these early alternative powered cars of both major types. But, as it turned out, they were also quite durable. There were still, even at this writing, quite a number of them extant. If they are earlier models, specifically from the mid-1910's on back, their usual styling markedly differed from gasoline cars. If they are later models, there was a general trend of more conservative styling appearing. One type quickly moved away from the other in styling.

As far as electrics, we have observed, "the style of the basic electric car didn't [sic] change over the years and became an anachronism in its own time." More pointedly, "Electric cars of 1920 looked exactly like their counterparts a decade earlier."[318]

This is a unique circumstance. Neither the gas car nor the steamer *ever* suffered this malady. In fact, the last two types increasingly began to resemble each other as time went by. Both had recourse to frequent appearance changes. In particular, the steamer, with less feminine virtue to appeal to, **had** to be more adaptable, much sooner.

The later steamers' frequent revisions to appearance, of course, was more than coincidence. Steam car makers of the early twenties—

Stanley and the Stanley clones of Coats, Delling & Company—were quick to realize the car-buying public of the era liked the look of the typical gas cars. That 'typical' gasoline car was already by then water-cooled, with vertical four-cycle engines situated up front under a clearly delineated hood, complete with shaft drive to the rear axle.[319] There is another indisputable assertion. Ultimately, those steamers which survived the longest looked the most like traditional gas cars.[320] The steamer was a powerful mode of transportation. But not all who looked at this type were taken in by it. After a couple of years experimenting with steam engines, pioneer Everett Cameron (1877-1965), for example, decided the, "car of the future would be propelled by an internal combustion engine."[321]

But, the steam car, like the electric, had its inherent disadvantages as well. Its biggest was the dreaded—and time-consuming—start-up procedure. The steam engine type had none of the complicated valve timings, or the moving parts of the four cycle gasoline motors, to say nothing of the complicated, unreliable ignition systems of the day. It also lacked the heavy batteries of the electrics.

These were all pluses. But, the process to start it was much more involved than its rivals. We need to briefly examine at this point just how entangled it could be.

At the start, one had to make certain the boiler and water tank were full. On the earliest steamers, which frequently lacked gauges, this often meant crawling underneath the car. Then, the fuel had to be pumped, by hand, to the minimum operating pressure. Then, the pilot burner had to be warmed up. This was accomplished through one of three rather unsavory methods: burning gasoline in a cup (now that sounds "safe!"); an alcohol wick; or an acetylene torch.[322] Then, the pilot had to burn until the vaporizer was hot, then one had to *carefully* open the main burner valve, while checking to make sure it was working properly. Got all that? This was a good time to oil the moving parts while the water was warming. A deed that needed to be done nearly as often as the car was run, we might add.

But, if the water level was not at proper level, an automatic check would shut off the fuel. Then our early steam connoisseur would have to retrace his steps back to the beginning. If the water level was too high, on the other hand, another device, the steam automatic, would also shut the unit down. Again, back to the drawing board. In short, steamers could not be driven until all the water had been worked out of the cylinders, and the entering steam had ceased to condense.

On the earliest steamers, including the Stanley and the White, preparation time could easily take up to 45 minutes.[323] And, none of this takes *routine* maintenance into account. The foregoing clearly showed the need for the driver of a steamer to possess both fine mechanical ability and lots of patience.[324] Not to mention lots of free time.

Almost as serious, winter time presented more problems. In those halcyon days before anti-freeze and more modern advances like snow tires came along, the pioneer car owner in colder climates typically drained their radiator and put their car up on blocks until warmer weather. Any offending liquid that could freeze in the car's system was usually relieved by this process. That was sufficient for internal combustion cars. If the gas car owner drove of a day in bitter weather, he would simply drain the car's radiator again that night.[325]

Owners of steam cars, on the other hand, could only drive their cars at great risk in the colder weather. Freezing pockets of water could put cracks in boilers, with very unpleasant consequences. Freezing could even occur *while* the vehicle was in use. But, in typical 'Catch 22' style, if the steamer were **not driven every day,** "fittings would corrode and packings dry out."[326]

There is another difficulty. To date, no good substitute for water has been found. Freon was tried by a few steam car adherents. In the late 1960's, William Powell Lear (1902-1978), who was most famous for making Lear jets, tried to build a new type of steam car. He tried Freon in lieu of just plain water. But this substance broke down at lower temperatures, and otherwise proved unsatisfactory. Lear then offered, "a million dollars for a working fluid that would bring about 40 per cent thermal efficiency."[327] At this writing, a reasonable cost liquid/fluid that fulfills this important requirement remained elusive.

If that were not bad enough, just getting suitable water could be a big problem in and of itself. Notice this insightful observation: "Especially in hilly areas where soft water in horse troughs was abundant, several American makes of steamers gained rapid and wide acceptance."[328] Moreover, water used in steam boilers *had* to be clean; otherwise, the intricate passages of the boiler system could clog up. In areas like the American Southwest, where *any* kind of water is often at a premium, water in horse troughs for steamers was never a practical option.[329] As for its gasoline competitor, let our old friend Maxim relate one of his experiences. The water in his gas car's radiator, "was horribly stagnant and covered with green slime. . . [but] it cooled as well as the purest spring water."[330]

Then, finally along these lines, we have a hoof-and mouth-disease appearing in the early years of the teens. Several places reacted by promptly doing away with public troughs. Of course, this was not a direct attack upon steam cars, but it was a serious blow for the type. On the other hand, it is hard to make a case that steamers were killed off as a result. The epidemic did not reach its zenith until around 1916; by this time, the steamers were well on their way out anyway.[331]

Steam car makers had an additional concern unique to their type. Some consumers lived in fear their boilers could, and every once in a while did, explode, often without provocation.[332] There is a big headache here: In order to be of practical use in automobiles, steam engines needed to be light and compact so steam pressure **had** to be correspondingly increased to wring sufficient motive power from the small motors. The high pressures that resulted then brought on the aforementioned fears. In fact, an apparently common contemporary bit of dry wit regarding steamers ran something like, "it was a toss up whether he [the driver] was going to travel straight ahead or straight up."[333]

This unfortunate circumstance, humor aside, really put steam cars at a disadvantage compared to the gas cars. There is a certain irony here. As one source succinctly put the matter, "Afraid that it [the steam car] might explode, people turned to cars with internal combustion engines which exploded many times a second."[334] True enough, but at least the internal combustion engines had controlled detonations taking place *inside* the engine. Not "blowing people up" outside the engine. This last one was an unpleasant image the steam car could never quite shake off. No matter how hard steam car companies tried.[335]

We should also note here, "It is not high pressure, but the size of the boiler. . ." that is important. "A huge industrial boiler operating at only fifty pounds of steam may be devastating if it lets go, while a flash boiler. . . at five hundred pounds can suffer nothing more than loss of steam in an explosion."[336]

Such calculated, reasonable logic, although unquestionably correct, had little apparent appeal for laymen. Such was some of the prejudice confronting, and confounding, early steam car proponents. The situation, from a more remote time, seems all blown out of proportion, if the reader will pardon the pun.[337]

Nevertheless, most people apparently did not relish the idea of sitting over a boiler with 500+ pounds of pressure per square inch, no matter how seemingly "safe." The consistently modest sales figures of the steamers after their initial popularity bears this out. Said one *1899* source, "When the word steam is used it naturally *brings to mind a certain uneasiness.*" [italics mine][338] As we see here, at least one early company sensed the concern, and tried to address it. Said the ad copy, buyers of, "Victor steam automobiles need have no anxiety for the boilers are tested and insured."[339] Early steam companies really needed to get the "good" word out about their product.[340]

XI. *PROMOTING THE PRODUCT.*

Yes, advertising was the one method by which the steam car companies sought to allay the public's anxiety. For example, an ad for the 1905 Grout Steam Car detailed it had a, "New and Improved 18-inch Boiler and Fuel System. . . [in which] Back Firing [is] Impossible."[341]

However, just bringing up the possibility of anything going wrong with the steamers was anathema to some people. White of Cleveland, unlike that other famous steam company, Stanley, liberally spent an advertising dollar or two. "Safety is the White keynote," says one ad. Follows a long, almost garrulous, detailing of the virtues of the car. Finally, with a nod towards gentility, "A lady need have no fear of soiling even a glove in running the car."[342] Madison Avenue, eat your heart out!

Framers of gas car ads, in sharp contrast, often very quickly moved to dispense unwarranted concerns out of fear of possible adverse publicity. Like simply failing to **mention** them. An ad for a 1905 Peerless Touring Car, for instance, neither mentions a safety term nor implies anything along this theme. "Peerless Direct

Drive Touring Cars are justly celebrated for the correctness of their mechanical construction," says the ad, simply and quite effectively.[343] And confidently.

The emphasis in this particular case on the positive, not on what could go wrong.

The electric manufacturers advertised, too. One of the best of the electric car makers in terms of its advertisements was the Columbus Buggy Company of Columbus, Ohio. One of its ads emphasized ease of operation, and low cost of maintenance, among other noteworthy features. This firm tried to accentuate the positive. Significantly, *not* mentioned were the batteries nor *their* associated drawbacks. However, "the absence of oil, smoke, dirt, and noise. . ." makes for, "all that you can possibly obtain in any motor vehicle."[344]

Nonetheless, all electric cars were having a difficult time by then trying to compete with the gas-engined cars. Even to survive. Furthermore, not all electric car firms ignored the strengths of the gas car's market position, to their peril. "Detroit [Electric Company] attempted to market an electric which looked much like a conventional gasoline car," quotes one source.[345]

This concern, unlike most of its competitors, did not avoid the unpleasantries associated with electric cars. An ad from 1910 has bold print at the head of the page boldly announcing, "These cars will carry you a whole day with current to spare." The caption really emphasized the range of the car on a single charge, day and night, farther, "than you will reasonably care to travel in one day."[346] That last statement was certainly open to interpretation, according to the person and the contemporary state of car developments. American society was already becoming more mobile than ever!

We have already surveyed who composed the target market. Significantly, a 1914 Detroit Electric ad boldly made clear its target market. "Only ladies and a child are shown in the drawing."[347] Thus, by implication, the electric was a 'lady's car.' Men apparently were more interested in its 'brutish, savage' cousin, the gas car. With a few hardy exceptions, like Mrs. Cuneo, the fairer sex disdained that choice.[348]

By necessity, electric manufacturers had to cater to women, for reasons already alluded to. Lavish, often full-color ads of the major electric firms frequently appeared in significant journals of the day like *The Ladies Home Journal* as well as *Country Life* and *Suburban Life*. As the alternative cars lost out in popularity, however, these same ads gradually disappeared.[349] By 1920, virtually no ads were issued for either major type of alternative cars; by then, not only had demand all but dried up, so had the advertising dollars of these firms.

In retrospect, any momentum the alternative cars *had* gained certainly exhausted itself quickly. Nonetheless, so soon as the production /registrations began their hopeless slide in favor of the gas cars, several alternative manufacturers did their utmost to react to this trend. Some concerns hastily switched over either partially or entirely to internal combustion engines, in a, 'if you can't beat 'em, join 'em,' type of attitude. White of Cleveland was among this group, abandoning steam production altogether after 1911 to concentrate solely on internal combustion vehicles.[350] Others, as we have seen, tried restyling to meet the enemy head-on, all while keeping faithful to alternative power. For a while, at least.

There were still other available means open. They were utilized early on. Even as early as 1903, the Woods Motor Vehicle Company of Chicago had offered an electric car, "made to look like a gasoline automobile by utilizing a false hood or bonnet."[351] But, in 1916, the same company was to be among the few to unveil an entirely distinct—and unusual—type of dual power car. It made use of a gas engine for sustained long runs, and of an electric motor for tooling around town.[352] The Krotz Gas-Electric, which we have already looked at, was another of this genre.

Other concerns that refused to adapt to the changing times, very quickly went by the wayside. The early era of experiment was nearing its end. But the underlying problem by the late teens went much deeper than that. A new wave of conservatism among buyers was sweeping across the automotive landscape.

Conformity was already appearing in the industry. The general trend was to coalesce towards that typical water-cooled gas car the consumers craved so much. The Franklin, which built only *air-cooled* gas cars throughout its existence, adopted a dummy radiator design of more realistic appearance than its previous design for the 1925 model year, complete with fake radiator cap.[353] Previously, the firm had bowed to the inevitable when it first half-heartedly put a "radiator" on the front of the 1921 models. Why had the Franklin factory made this latest change? Sales resistance had caused a group of dealers in the summer of 1923 to petition the factory vigorously for a more conventional looking product.[354] Did the stratagem work? Yes, but it was to be only a temporary solution.[355]

The foregoing may help explain why the unconventional alternative cars could not long survive in the prevailing business climate. Especially if unusual gas cars were themselves encountering sales resistance. Despite having similar appearance features. Nor were these the *only* concerns.

XII. *STEAMERS AND ELECTRICS GO RACING.*

Along with the misconceptions about boiler explosions, a number of other fallacies have arisen about the alternatives. For example, considering what we have already looked at, some might wonder if the steamers in particular ran at all. Actually, they often managed quite well, Thank You. Most of the earliest land speed records, in fact, were held—or broken—by steamers. A goodly portion of these record holders were Stanley Steamers.

Much has been recorded about the eccentric Stanley brothers and their famous steamer. Enough so we might confine ourselves here to the oft mentioned, but little understood, effort to break the world's land speed record. It, too, had been shrouded in mystery to this writing. Since *much* misinformation about this topic has been published over the years, we need to extract the facts from the story. We have one factual story and one of the same ilk as the Belle-Isle tale to relate.

For starters, we have accounts of a Stanley Steamer breaking the land speed record once upon a time. The place was at Ormond Beach, not far from Daytona Beach (the same strip was commonly called just

Daytona Beach at this writing), Florida, on January 26th, 1906. The recorded speed on the occasion was 127.66 miles per hour.[356] That is a fact.

The boiler of this racer boasted a steam pressure of 800 to 900 pounds per square inch (psi), and had a compact two-cylinder steam engine with 4.5 inch bore X 6.5 inch stroke, for just 207 cubic inches of swept volume.[357] Of course, this engine was double-acting.[358] The Stanley did not employ a transmission, in common with most steam cars. But neither was this particular car geared direct one to one. The racer was instead geared at 1:1.75.[359] It was rated at 20 hp.[360]

The driver of this beast, Fred Marriott (1872-1956), was the *first* human ever to travel more than two miles in a minute by any means.[361] No gasoline-engined car had as yet even approached that rate of speed. In fact, Louis Chevrolet (1878-1941) was also present that day at the beach.[362] His "ride," a 200 horsepower French-made Darracq, featured a mammoth overhead valve V8 gas engine, and is an excellent example of our 'age of the giants.' This car boasted many of the latest advances from the Continent. The engine dimensions were certainly exceptional in and of themselves, with pistons of 6.7 inch bore X 5.5 inch stroke. This figures out to exactly 1551.4 cubic inches; it thus displaced almost 7 ½ times more cubes than its formidable steam rival.[363]

But the best time of the day's events for Chevrolet was one run of 115.2 miles per hour, which, nevertheless, still easily set a gas car record. The Stanley's other chief rivals that day were a 110 horsepower Fiat and an 80 horsepower Napier, also with internal combustion engines. All three of the European entries were highly prized, but they could not match the power of the little steamer. Marriott not only handily won the meet, "but even more gratifying was proving the superiority of the steamer over his highly publicized rivals."[364]

Obviously, the little Stanley engine was producing much more horsepower than its gigantic gasoline rivals, but there was a catch. All factors being equal, it would literally run out of steam long before any of its gas-engined competitors needed a refuel. As a demonstration of this, Marriott swept every event in the meet, except for the final event. In this one, his valiant car was pitted against two Fiats in a five-mile grudge race. With the opportunity to really put the gas cars in their place, the steamer's nemesis instead reared its ugly head. Marriott's car did indeed demonstrate the steam car's big weakness of lack of range. It seems the car ran out of steam, literally! He came in a distant, disappointing third in that final competition.[365]

Word of Marriott's feat of humbling much larger gas-engined rivals nevertheless spread quickly.[366] The performance image of the Stanley was even the stuff of legends for a time following the achievement. As Musselman wrote, "Almost every one of my youthful acquaintances believed there was no limit to the speed of a Stanley Steamer."

In a similar vein, was this bit of youthful exuberance, "for in one minute, with the throttle wide open, it was thought that the car might easily take off into the sky."[367] The rumor was even bandied about for a while that Stanley Steamers had *no* top speed.[368] What is more, "The production 1907 Gentleman's Speedy Roadster could reach over 75 miles-per-hour."[369]

If *this* were a real historical performance, though, it pales into insignificance compared with what followed. When Marriott returned the following year to Ormond Beach to vindicate himself, a much greater legend was to be born. One of a mixture of reality *and* of fantasy.

First, the facts. For this second attempt, the Stanley brothers apparently had improved the car. Steam pressure had been boosted again, this time, "up to 1[,]300 pounds per square inch."[370] With other refinements, the car was said to be developing 250 horsepower (remember this was in 1907!). It is worth noting the car had not been tested prior to Marriott's record-breaking attempt.[371]

Now the issue gets cloudy. Eyewitness accounts of what ensued are, at best, confusing. At the worst, misleading.

Francis E[dgar] Stanley (1849-1918) was present that day, January 23rd. He said about the racer, it crossed the starting line, "at a rate of speed never before seen."[372] Then, disaster. In Stein's words, "he [Marriott] hit a bump on the sand, [and] became airborne."[373] The boiler, "shot down the beach for a mile."[374]

Now, to the controversy. Marriott, who was nearly killed, later claimed his speedometer was reading over 190 mph and still climbing. This exceptional rate of speed has been disputed by some reputable sources. Unfortunately, from the historian's point of view, it has also been perpetuated by a far greater number of other accounts.[375]

Rumor mill, indeed, said the speedometer pegged as high as 197 mph.[376] Dr. Robert U. Ayres, then [1967] vice-president of International Research and Technology in Washington, D. C., said the terminal speed was exaggerated. After looking over the situation, he wrote it was, "probably impossible for the Stanley engine to go that fast."[377] *Standard Catalog* puts the terminal speed at a more conservative, but believable, "over 150 mph."[378]

At this opportune moment, we should mention a few pertinent facts: 1.) According to Marriott's own account, he fails to mention looking at a speedometer; 2.) Also, as Ray Stanley (1894-1985), son of F. E. Stanley, so deliciously put the matter, "It is even questionable whether speedometers were made in 1907 that would accurately register speeds as high as 197 mph."[379]

As for further racing efforts afterwards, Francis Stanley stated succinctly, "We decided never again to risk the life of a courageous man for such a small return."[380] A factory sanctioned race never occurred again.

Even earlier, in 1905, Louis S. Ross (1877-1927)—who would later design his own steamer—in a Stanley, "with a remarkably streamlined body [had] pushed the record for the flying mile to 94.73 mph."[381] The Ross machine, a.k.a. Stanley "Teakettle" accomplished that feat in 38 seconds using a 600 psi boiler.[382]

Nor were steam car makers alone interested in displaying the power of their machines to a still credulous public. We would be remiss if we did not briefly mention a few additional racing facts, these concerning electric cars.[383] In fact, there has been much less controversy surrounding the latter alternatives. On April 29, 1899, near Paris, France, the famous early race driver Camille Jenatzy (1868-1913) piloted his *La Jamais Contente* ('The Never Satisfied'), an oddly-shaped electric beast, to a one-way best time of 65.8 mph in the flying kilometer.[384]

The problem was, Jenatzy virtually destroyed his car's very

expensive battery to achieve this noteworthy feat. In fact, after just the one pass, "it is doubtful whether Mr. Jenatzy had enough juice left in his battery for even one more kilometer."[385]

This points out a major disadvantage of the electric cars they were never able to overcome: If driven hard, their already limited range diminished still further. And quite dramatically, we might add. Said Kenneth Purdy, who drove a few, "Almost any good electric of the dozens that were made. . . would do 60 miles per hour—for a couple of blocks."[386]

On Memorial Day, 1902, Walter Baker, at the wheel of one of his specially constructed Baker electrics, made an effort to cover a mile in just 40 seconds on Long Island. A first run of 78 mph was unofficially obtained; then a freak accident plunged, "the racer into a crowd of spectators. One [person] was killed and four or five others seriously injured, one fatally."[387] The vehicle in question, dubbed the *Torpedo,* was doing about 120 mph at that point, while utilizing only one-half of its 40-cell battery pack. What was its maximum speed with all cells working?   Frankly, "No one ever volunteered to drive the car to its full capacity."[388]

The *Torpedo* had an even more pronounced shape than the racing Stanleys or "Whistling Billy," a similar racer from the White company. Not only was this battery arrangement heavy, it was inordinately expensive. But Baker, in trying to prove the electrics could be competitive, was also sparing no expense here. In fact, economy never factored into the equation. Thus the *Torpedo*, with its expensive drivetrain, cost $10,000 just to build. That sum of money in 1902 would have allowed for a princely annual allowance.[389] But this expensive car did indeed point the way to victory.

The pioneering electric had been there even earlier. Frequently, in fact. In the first track race ever held in the United States, on September 7th, 1896, an electric had pointed the way to victory. The setting was at Narrangansett, Rhode Island; close enough for the "Cream" of Society to be present in formidable numbers. There were only seven cars involved. Five of those were Duryea gas cars, along with an "Electrobat"[390] and a Riker Electric. The latter won all five runs—"something that would have been impossible for an electric over any reasonably long distance."[391]

The White company was interested in proving the performance potential of its cars. Rollin Henry White (1872-1962), who graduated from Cornell University in 1984—his graduating thesis dealt with the gasoline engine and its application to 'horseless carriages'—ultimately decided internal combustion cars were the future. But not before a decade-long stint with steam cars. The company fielded a great racer in the meantime, the "Whistling Billy." This rather quaint name stemmed from the sound of the racer's rather high pitched exhaust.[392]

The odd shaped racer was powered by a two cylinder compound engine.[393] The vehicle made its racing debut on May 20th, 1905, at Morris Park, New York. There, "it completed an exhibition mile in 53 seconds to equal a track record."[394] Webb Jay, a celebrated race driver, at the wheel of his trusty steed, set a new world's record of 48 4/5 seconds in the mile (speed of 73.75 mph) at Morris Park, July 4th, 1905.[395] On this occasion, the White was at its best in vanquishing two gas cars: a 60 hp Thomas[396] and a 90 hp Fiat. The latter featured an enormous 995 cubic inch T-Head (7.09 in. bore X 6.3 in. stroke). As Clymer states, "When one considers this record. . . was established only five or six years after the average American had glimpsed his first automobile, it looms as an accomplishment of real proportions."[397] In fact, the White racing efforts had inspired the Stanleys to proceed with a program of their own.

Nor was this all. On August 11th, 1904, at the World's Fair Races in St. Louis, Jay beat a whole field of gasoline-powered competitors with his 10 hp White. All of the competitors had 24 hp engines: a 227 cubic inch T-Head four Peerless; a Pope-Toledo (T-Head four cylinder of 4.25 in. Bore X 5.25 in. Stroke for 298 cubes), a three cylinder St. Louis Tonneau and a Panhard.

Not that steamers always got the upper hand. On July 9th, 1905, less than a week after Webb's record setter, that stalwart gentleman squared off in a grudge match between his trusty steamer and two gas cars. All three were Cleveland-built cars. The event was staged at "Hamlin Park," sponsored by the local St. Paul Automobile Club in Minnesota. The gas entries were both packing big engines. One, a Peerless four cylinder racer (T-Head engine of 679 cubic inches—6 in. Bore X 6 in. Stroke) boasted 60 hp. The other was a Winton of 60 hp nicknamed the "Bullet." On this occasion, the great Barney Oldfield (1878-1946), in the Peerless 'Green Dragon' was a non-factor, but still ended up in second when, at the end of, "the third lap[,] difficulties with the White steamer caused Webb. . . 'to run out of steam.'" This in spite of the fact, "[Earl] Kiser [in the Winton] and Webb fought a tight race for the first three of the five laps."[398] Kiser, whose L-Head four cylinder engine displaced 792 cubic inches (6 in. Bore X 7 in. Stroke), won out that day.

In short, there were many performance events participated in by alternative cars; far more than we want to survey here. And why not? Road racing in those primitive days was a good testing laboratory. "This constituted a rigorous form of road testing. . . [proving] something important had been accomplished—in this case, before thousands of witnesses."[399] However, the barrier of lack of range all too frequently demonstrated itself in these early racing contests.

The country buzzed with all the latest developments in the field of motoring. The speeds were entirely new to a generation inured on horse and buggy, and frightening to some. Some doctors, in fact, even insisted—to the bemusement of more modern drivers no doubt—this whole racing practice could be medically dangerous. They were not sure if humans, especially the more genteel ladies, could stand to travel at such high rates of speed.[400]

There was even talk about limiting the power outputs of cars. As early as 1901, the year millionaire tycoon William Kissam Vanderbilt, Jr. (1878-1944) won a 10-mile championship race in Newport, Rhode Island, one journal editor was seriously asking, "It may well be asked if the limit of speed in racing vehicles has been reached."[401] His speed was a sizzling 39 mph.

XIII. *FAR-SIGHTED PIONEERS SEE THE END OF THE ROAD FOR THE ALTERNATIVES.*

There were quite a number of influential entrepreneurs in the early days. Some were more far-sighted than others. The same

Locomobile company that made the early steamers switched to internal combustion cars. This was after making about 5000 steamers. Andrew Lawrence Riker (1868-1930), who was running the company, thought he saw the end of the road coming for the alternatives. Early in his career, Riker had believed in the electric. His car had been the one we have already observed winning the first track race.[402] The thinking money might have stayed with electrics. But the talented engineer joined Locomobile in 1902, just about the same time as that company's steamer was needing an update.

At first, he designed a gas car to supplement the steamers. Soon, "Riker's [new] car, with pressed steel chassis, proved so attractive that by 1903 Locomobile were out of steam production altogether."[403] The decision would prove to be wise. In the short run, it did not appear so. In that same year of 1902, Locomobile made 2750 steamers,[404] 250 more units than the Curved Dash Olds, the next best-seller in the country. It must have seemed an awful gamble to mess with a winner!

Nonetheless, the company was committed. Riker, who had always had a bend towards racing, then set about creating the perfect racing gas car, *circa* 1906. The heart of the new racer was a behemoth F-Head four cylinder that was fully 42 inches long, with massive cylinders surrounded by sheet copper for water jackets; a concession to weight-saving. Like the Stanley Beetle, this Locomobile (the "Old Sixteen") was specifically designed for the 1906 Vanderbilt Cup Race.

Not only did this racer compete in the Vanderbilt that year, it won. The first American car to do so.[405] The Old Sixteen was yet another of those monster internal combustion engines we have been looking at. The bore of each cylinder was 7.25 inches, with a stroke of 6.25, for 1032 cubic inches.

When we compare the dimensions of this particular car with the Stanley racer we looked at before, we can see some of the steamer's problems. A wheelbase of 110 inches and a curb weight of 2204 pounds on the Locomobile compared favorably with the Stanley. The latter used a 100-inch wheelbase, but weighed nearly the same at 2195 pounds. It was clear, from looking at the two side-by-side, the Locomobile was the more conventional—as well as better-looking—of the two.

Nor was the playing field all that level, even then.

The steamers really had the problem of range. They could start, but could they finish a race?

Fred J. Wagner, an early racing car driver/official, said about many of those early contests involving steamers, "There were moments, however[,] when I feared I'd [sic] have nothing to finish, for the steam died down as the contenders hit the back stretch, and snails were fast by comparison."[406]

One more example might suffice to demonstrate the alternative car's general lack of range. In the 1895 *Chicago Herald* race, the participants covered the approximately 52.4 mile round trip from the Jackson Park in downtown Chicago to Evanston in conditions that were, well, primitive.

The run was set for Thanksgiving Day, November 28th. There happened to be substantial snow on the ground that day. Two American electrics participated, the rest of the field had internal combustion gas engines. One of the two alternatives, a Morrison-Sturges Electric, soon patently displayed it was out of its element. "Driving through the deep snow overworked the [vehicle's] battery. . . compelling frequent stops to prevent the motor from burning out."[407] The gesture was futile! Although the participants did their best, that unfortunate entry was finished before noon.

The other electric, an Electrobat II, from our old friends Morris and Salom, lasted only eleven grueling miles of the course. It failed even to make the first relay station, and was simply driven back to the factory, which fortunately was near by. This entry apparently used up two complete battery changes in very short order.[408]

An American Duryea ultimately won the race. Notably, it had a gasoline engine.[409] Steamers were conspicuous by their complete absence from this important milestone.

Also, said Peter Helck, "Because the backers of the electric buggies had not provided recharging facilities along the route, they were viewed not as genuine contenders but rather as publicity-seeking opportunists."[410] Such they proved to be!

To the point, one of the few constants of those early days was the fact any long-range endurance run was participated in *only* by gasoline cars. With good reason. Neither the electric nor the steamer could handle the requirements.[411]

XIV. ELECTRICS HAVE SOME ADVANTAGES.

Despite the above, the electrics had some advantages. Factors unique to their type. They were "silent," some insisted; i.e., less noisy, and could be operated, others insisted, by a mere child. For certain, drivers of this particular type did not need the strength to crank the moody gas engine nor the patience/mechanical ability to warm up and run the steamer. To start the electric car, all one had to do was turn a switch, or press a button. Or, just work a lever one way or the other. Of course, ready access to facilities for recharging and changing batteries when they ran low was also necessary. Which was quite frequently, we might add.

Electric car makers, nonetheless, were justifiably proud of the ease of control of the type. Buffalo Electric, for instance, likened the guidance of its models to walking. Both actions required the same movement of the feet. To start, "you raise one foot, to stop you place both feet on the ground."[412] Thus, just a single foot pedal operated the car.

Contrast this with the other major types, both of which utilized a bewildering array of gauges and controls. Early steamers in particular, in every photo and/or description available, were just so. Some of these gauges were located at what passed for a dash, others were under the car, or even somewhere in between.

And, it must needs be said electric cars were by far the most reliable of the three early buggy types.[413] Hence its nearly universal appeal to the female gender. Notably, the electric car seems to have become forever associated with that brief period in which it flourished, though; from the late 1890's to about 1910. The fashionable debutante of that period would scarcely go out without her, "rainy-daisy skirts three inches above the ground, [and] armor-plated corsets."[414] If she drove at all, her car was almost invariably electric, although the youngest of the set would live to see the day when many people questioned openly whether there had *ever* been any motive force for cars other than the internal combustion engine.[415]

In fact, the pages of *The Horseless Age* questioned whether the electrics could even compete with the presumably outmoded horse. "In the present [1899] state of the storage battery art, electricity cannot [sic] compete with horses in public cab service, but is of necessity confined to luxurious use."[416] In this case, the 'luxurious use' was confined to the fairer sex.

As it was, not everyone was blind to the virtues of the electric. In its heyday, a surprising number of people owned electric cars. Among them was the distinguished Mr. Henry Ford (1863-1947). Yes, the same man who pioneered the inexpensive car for the common man, the unassuming gasoline-powered Model T, once owned a Model 43 Detroit Electric. He both frequently used and apparently preferred it in deference to any gas car.[417] This presumably included his own.

Even Edison, the same man who did not see a long-term electric solution, had the foresight to purchase the second Studebaker Electric ever built in 1902. So it seems the electrics must have had *some* merit after all. He kept this car as his own ride for 23 years; a remarkable length of time for that day.[418]

There is another factor that appealed to many early motorists, especially women. Electrics were really the quietest of the three main types. The steamers were nearly as silent, but they were not complete without their plumes of oily smoke. Women of quality usually did not appreciate *this* feature of the steam car. Oil stains on rainy-daisy skirts apparently were none too attractive. At first, this smoke posed no concern. By the late teens, though, this drawback was becoming a real nuisance as the number of cars on the roads rapidly increased.[419] But, ah!, the steamer was a silent navigator. Some even complained, for instance, the Stanley Steamer was probably too quiet. "Unalert pedestrians wouldn't [sic] hear it coming and might be struck down."[420]

Speaking of "quiet," one must hear those early gasoline motors "run" to appreciate the improved internal combustion models that would come later. Yes, the crude, less "civilized" gas car had no problem announcing its presence. Early eyewitness accounts are replete with such references. "The early [gasoline] cars needed no noisemakers to warn people of their approach," said one source laconically.[421] So horns appeared last of all on the gas cars as a matter of course. Moreover, this was a major drawback to the pioneer gasoline buggies.

The gas car did have one more major disadvantage: unlike the steamers and electrics, gas engines needed multiplication of torque to work, better known as the changing of gears. Transmission shift at the very beginning, while necessary, was inordinately complicated.

In 1902, Packard introduced the soon-to-be-famous "H" gearshift pattern.[422]   Still, transmissions lacked true synchronization until the adoption of synchromesh by General Motors in 1928. Until then, all transmissions required double-clutching. This meant one had to depress the clutch pedal once to engage the clutch and a second time immediately after to change to the selected gear.

In 1896, Count Albert de Dion (1856-1946) and George Bouton— two highly influential European connoisseurs—developed a thin synchronizer that worked in relatively slow turning mechanisms, but could not be used at high speed. In 1927, Peter Salerni designed a dog clutch which did not allow the gear teeth to re-engage until they were, "running at the same speed." In 1929, Bernard Thompson combined these two closely related features, making double-clutching in cars a thing of the past.[423]

There *was* a modicum of complication in all of this, although it certainly pales by comparison with the disadvantages of the alternative cars. Flink states, "Before the introduction of synchromesh. . . shifting without clashing gears posed a formidable problem for the average driver." And Maxim wrote of the singular difficulties of coordinating the clutch pedal and the shifter.[424] Moreover, the barrier of the Self-Starter was not the only obstacle for early women motorists in the gas cars. George Robertson, who drove Old Sixteen to victory in

1908, wrote about the beast, "it took an ox to turn the wheel. There were a hundred and ten pounds of pressure on the clutch pedal." After one racing experience of several hours' duration in pushing down and engaging the clutch with his left foot, he related, "you walked sideways."[425] Now, admittedly, this was a race car, but the disadvantages even in ordinary gassers was obvious. Those same extra large engines we have been covering produced loads of torque and required 'King Kong' clutch effort. This really showed what those early motorists were up against; with 'nary an automatic transmission in sight.'

Still, we can here recapitulate on one of the major disadvantages of the alternative cars: Their excessive weight in relation to the power provided by their engine or motor. This turned out to be a basic issue of both weight and bulk. David Cohn (1896-1960) quoted the following figures for power-to-weight ratios: 371, 840, and 185 for steam, electric, and gasoline cars, respectively. He summarily concluded, "A great deal of the power developed by the electrics and steamers was devoted to. . . hauling its [sic] unwieldy bulk."[426] This was one of the fatal flaws.

To put that in perspective, notice the following statistic: Operating radius of the three types, assuming everything else were equal, "the steam car would develop nine horsepower hours, the electric fifteen . . . [and] the gasoline car forty horsepower hours."[427] In lay terms, "the same amount of technical effort would inevitably produce better results with the gasoline car than the steamer."[428]

On the surface of the matter, the steamer seemed to have the upper hand. Steam as a motive force for engines came out long before the American Civil War (1861-1865), and had powered many naval vessels by the late nineteenth century. If anything was going to supplant the horse, it seemed like the steamer should be it. Even by natural progression. After all, locomotives were steam powered, and *most* of the earliest attempts to power cars had been with steam. It does seem curious that steam locomotives were on the scene for half a century before the first serious attempts to power cars were made. The question can be asked, then, 'Why did this process take so long?' The answer was an obvious need for a practical fuel. Something besides coal or wood. Something more portable, less space wasting. Something which did not take up most of the available space just to haul it.

There was a solution. In France, pioneer Leon Serpollet (1858-1907) had, "as early as 1885" designed a very efficient flash boiler with a portable fuel. Entering water was almost instantaneously converted into vapor inside the confines of a coil of this design, which meant water was heated only as needed. This system was thus not only more economical, but lighter in weight because of its more compact components. Additionally, it was, "safe [with limited water in the steam generator]. . . and the possibility of an explosion was remote."[429] In 1890, Serpollet made use of a liquid fuel, one of the very first to do this. This was paraffin wax.[430] Although in the beginning Serpollet's cars could only reach a maximum of seven miles per hour, improvements came quickly. In fact, in 1902, using one of his "racers," Serpollet made history. He handily captured what was then a new land speed record: 75.06 mph.[431] With this impressive array of achievements, it is very disheartening to report on the sad end of the inventor's career.

After securing the backing of wealthy businessman Armand Peugeot (1849-1915), Serpollet appeared to be set for a long run. But this partnership was of short duration. In 1891, Peugeot, fascinated with the newly developing gasoline engines, summarily withdrew his funding from Serpollet. The latter had no interest in internal combustion designs. As Serpollet lacked the money on his own to seriously pursue his designs, this effectually meant the end of further experiments for the unfortunate discoverer. The gas engine had claimed another victim.

Moreover, this was fairly consistent throughout the early days. In engineering terms, the thermal efficiency of steam engines was very poor compared to their gas engined-rivals in the same size range. The former had, "the great disadvantage of loss of energy due to inherent problems of heat transfer and storage."[432] Just as importantly, this factor would change little over time. Even the mighty Doble, which could be considered the *'Beau Ideal'* of steamers, as impressive as it really was, still had inherent problems. "The fuel consumption of under 10 miles to the gallon. . . points to a much less impressive level of specific efficiency."[433]

XV. WERE THEY GIVEN A 'RAW DEAL?'

There might still be lingering doubts the alternative cars were axed even when shown to be superior technologies. This is the third of the writer's tenets; that the alternative cars, despite claims to the contrary, were inherently *inferior* to the internal combustion car. The argument has been postulated the early steamers, for instance, were killed prematurely. Before their time; before they could adequately *demonstrate* their's was a superior technology.

To prove this was not the case at all, we can look very briefly at the checkered career of Abner Doble (1890-1961). Unlike the Stanley brothers, who never tried to offer a 'perfect' version of their car, Doble was both a steam car proponent and a perfectionist. He built few cars he was such a perfectionist; even these few did not satisfy his lofty aspirations. As an example, he claimed to have perfected a condenser of 95 % efficiency, which boasted a range of up to 800 miles on a single tank of water.[434]

In his efforts to build as good a steamer as possible, Doble, "filled them [the few he built] with thermostats, electrical switches. . . that made them extremely complicated and expensive."[435] All steamers were complicated, we must admit. But the Doble would prove both exceptionally difficult to handle and maintain.

And the initial cost was a real problem. This make's price, at best, bordered on the stratosphere. The Model E Doble *chassis* in 1930, which rode a very long 142-inch wheelbase, cost $6800 and weighed 4250 pounds.[436] And one would *still* require a body to mount on the bare chassis. Make no mistake! This immediately put the Doble into very elite, exclusive territory. Comparisons really bear this out.

A point of reference. A 1930 Cadillac V16 two-door convertible coupe, a complete car, cost just $100 more than the Doble chassis.[437] This well-equipped luxury car was the very epitome of extravagance in the depths of the Depression. And there were no Henry Fords of steam just itching to mass produce a steam car the average man could afford. Ever!

However, Doble's cars may be considered the ultimate steamers.[438] He designed a spinning burner connected to an electric blower, which put air into the gasoline/kerosene. This mixture was then fired by a spark plug, with the blower continuing to revolve while the fuel mixture burned. Not until maximum operating pressure were reached would the unit shut down.[439] Yes, there was potential.[440]

But Doble's total production amounted to just 42 cars between 1912 and 1930.[441] He was able to build a steamer—at the end of his career—that could be ready to run in less than a minute, with an improved main burner which required a mere two seconds to ignite.[442] In short, Doble was able to build a steamer that took away many of the pet objections people had always held against steamers. His cars were coveted by many wealthy/influential people.

But, he did it far too late, in far too small a quantity, and for the wrong end of the market to boot. If Doble had bowed to mass production, if he had lowered his lofty standards a bit. If, if, if. Counter factual history, again. In more sober retrospect, Doble's rather radical ideas probably came *too* late to save the steamer, just as Serpollet's were much too *early* in their own right.

The Doble's high cost betrayed another problem common to the alternative cars: Price. They generally cost more, much more, than comparable gas cars. This was inevitable. As serious steam people like the Stanleys, who wanted at least a competitive product, tried to keep pace, they had to offer something. Like new features and better quality. This forced them to raise prices.[443]

For instance, Stanley Steamers adopted neither condensers nor flash boilers until after Cadillac introduced its self-starter. This forced the brothers to act.[444] On the other hand, the Stanleys still suffered because of an old-fashioned boiler type. In fact, the early Stanley Steamers consumed water at the shocking rate of approximately one gallon per mile.[445] So, they, "ran beautifully but used a lot of steam. [The improvements] did not help in achieving a cruising speed high enough for 1925 and later."[446]

At last, even those few customers willing to put up with the numerous shortcomings were not to be found in numbers needed to sustain production. On the other hand, steam car makers like the Stanleys were not completely numb to the needs of these same consumers. We have already observed the cars were designed to run on gasoline from the very beginning. The brothers actually touted this as an outright advantage as the number of I. C. cars on the roads increased and the fuel became easier to obtain.[447]

Ultimately, the success of the Model T and others like it flooded the market with internal combustion cars. Soon, this was to be the accepted type. By the end of the century's first decade, steamers were relegated to just curiosities, and nothing more. It was all a vicious cycle. If people did not buy steam cars, positive stories about them did not get circulated. The buying public, it followed, generally did not know about the type. Nor did they want to!

The gas cars flooded the market so swiftly they became the accepted, 'normal' type. Said Stanley K. Yost (1924-   ) people who favored steamers would, "likely be branded a 'square,' an eccentric." Those later steam connoisseurs were, "victim[s] of 'antique-automobia.'"[448] All of this despite the fact a final 1925 Model F was, "capable of traveling at 150 mph."[449] The tradeoffs just did not balance out.

But, the steamers did have one final advantage. They were very durable. "A steam engine turns so slowly. . . and operates under such low stresses that almost nothing can harm it."[450] Unfortunately, this singular fact meant little repeat business by consumers as steam car customers.

As a final cap, around 1925, just as the Stanley firm was marching off into the automotive sunset, the steam engine made a proverbial last stand. The Stumpf Una-Flow Engine Company, of Syracuse, New York, tried to develop a more efficient steam engine for automobiles. The firm tried every trick it knew to wring more power and efficiency from the willing beast. This was a nice effort, but ultimately, "without appreciable success."[451]

In more modern times, there have been other bonafide attempts to revive the steamer. In 1968, the aforementioned William Lear commenced a project to develop the, "ultimate steam car engine." Later in 1971, the enthusiastic, and very premature, aviator bubbled effervescently, "within ten years the internal combustion engine will be a collector's item." One would then have looked for continuing, optimistic reports issuing from the same source. But, in 1974—while the first Arab Oil Embargo was in full bloom—and having spent some $17 million of his own money on various steam experimental projects, Lear quiely dropped the, "ultimate steamer."[452] A closer look reveals some interesting details. "Experimental automobiles used a six-cylinder engine in a delta arrangement," among others.[453] The inventor's scope knew few bounds, on paper. There were very exotic 6 cylinder chamber designs with 12 pistons, utilizing three crankshafts. The planned hp was, "760 for the racer and 400 hp for a passenger car."[454] There was lots of work with turbines. While he labored away, Lear consigned the internal combustion engine to the Smithsonian Institute with a chuckle. Alas!, like so many before and after him, the crafty businessman was doomed to disappointment.

This is not to say there were not a few notable features. The "crank" in a conventional engine was replaced by a cam. This latter, a doubly-inverted piece of helix shaped metal, acted as a distributor of power in lieu of a "normal" crankshaft. Improvements were also noted in quicker, more usable water temperatures. This could help to make the steam car more practical. The best part: "Fifteen seconds after depressing the push-button igniting the burners the vehicle can start to move."[455]   In sum, Lear's car was probably the steamer's best hope after Doble.

There was one more significant, postwar, attempt to revive the type. Robert Paxton McCulloch decided, in the early 1950's, to add a steam car to his company's line of small two-cycle engines and chainsaws. To help expedite this project, he brought on board none other than Abner Doble for the project. The latter had been in Great Britain, trying to interest British speculators in new steam cars. Between them, the pair managed to produce a solid working prototype of a very efficient compound engine, that featured a mere 20-second start-up time. Everything seemed rosy.

McCulloch was advised against going into production, however, as costs were already prohibitive and direct competition with Detroit's Big Three automakers could conceivably have destroyed the whole firm. Probably a pragmatic summation after all, but technologically speaking, a bitter disappointment. In 1954, this steam car engine died aborning.[456]

XVI.*WHY?*

Finally, we reach the ultimate question: Why? Why did the auto industry take a course that doomed the steamers and electrics to early extinction?

The explanations as to the reason the auto industry took this specific path it did have tended to follow two quite divergent lines of thought, both of which were "determined" by the basic hardware of the car. They are: 1.) Technological Determinism; and, 2.) Social Determinism. It has been argued by proponents of the first view the gas car triumphed because its design was superior and, thus, the alternative cars were weeded out as a natural consequence. That this action was inevitable. James J. Flink is prominent among this group.

The second line of thought, Social Determinism, asserts basically, the auto as it existed at this writing was due to the gasoline engine being selected just *because* it was preferred, not because it was necessarily better. It would appear the two are not mutually exclusive. The preferred internal combustion car unquestionably benefitted from better development time/money than the less desirable alternative types. The gas cars thus enjoyed advantages denied outright to the others.

The natural inference from this would be the gas car triumphed because of **both** influences. It must not be forgotten in this mix that the actual cars were tested on the streets of everyday America. The American drivers had accepted what they wanted, and ruefully rejected the rest. However, since the gas-engined car was clearly preferred, even before the perfection of the Kettering Self-Starter, this would lend credence to the Social Determinism argument. We must not forget, in fact, long before the self-starter was perfected, the public had selected the "gassers." The rapid, unceremonious decline of alternative cars in the National Auto Show bears this out.[457] There is another factor. For a long time, portable gasoline engines had been used on America's farms to perform a variety of duties. So the stretch from there to internal combustion engines in automobiles. This needs to be thrown into the equation by those sources attesting the gas engine was an unfamiliar concept to farmers, which could not be further from the truth.[458] There was another, little mentioned, factor in the grand scheme.

The constant tinkering of early racers with gasoline engines also helped the endear the internal combustion cars to the public. Especially among the wealthy young racing breed of the pioneer era. It has been postulated these rich, otherwise idle, young men, sons of industry and commerce, used the gasoline car to advance themselves in society. It was certain, all other factors being equal, that wealthy people were the only ones that *could* utilize racing machines. Cars were very expensive, as we have seen in our few examples.[459]

Having said this, there a couple of other good reasons why the gas car triumphed: 1.) On January 10th, 1901, a huge oil gusher called Spindletop—near Beaumont, Texas—erupted. Immediately, the supply of petroleum produced a great glut on the market. Crude prices soon dipped below five cents a barrel. The resultant manifold increase in the supply of gasoline, and a consequent drop in its cost, encouraged tinkerers of gasoline engines to continue working and refining the internal combustion design.[460] 2.) A fire swept through the Olds plant on March 9th, 1901. Ransom Eli Olds (1864-1950) went back into production with a single model, sourced with parts from a variety of outside suppliers.[461]

This unprecedented move by Olds immediately put those contractors into the car business. And into the end of the business that specialized in gasoline engines. Names like John Dodge (1864-1920), Horace Dodge (1868-1940), Fred J. Fisher (1878-1941)(of Fisher Body), and Alanson Partridge Brush (1878-1952). All of these, and others, entered the auto industry at the internal combustion end of the business. In point of fact, *all* of the early mass production cars were of the internal combustion variety. That in and of itself should tip us off to the gas car's inherent superiority. There was more.

Although not necessarily an accurate barometer of American tastes, European drivers embraced the internal combustion idea even more quickly than their U. S. counterparts. After an early vogue in Europe, electrics simply faded away from the popular perception. As for its steam driven cousin, "The steam car played a much greater role in the motoring scene across the Atlantic than it ever did in Europe."[462]

Henry Ford, as we have seen, contributed his stake to the coffins of the alternative cars. By offering a low-cost gas car, at a critical stage in car developers, he dealt a decisive, ultimately fatal blow to the pricier steamers and electrics. When the ubiquitous Model T hit the streets in 1908, the gas car became overnight the car of choice for most Americans. It was a natural fit. The Model T's versatility and simplicity of design quickly endeared it to American consumers. Neither of these factors would have been possible with alternative cars. In fact, "After Henry Ford introduced his inexpensive Model T, a full range of cheap gasoline cars hit the streets."[463] Even most of these cars cost less than the pricey steamers. Of course, this still further diluted the potential market for the alternative cars.

In addition, the times, well, they, 'were a-changing.' The steamer had thrived on the bad conditions of the early roads, when a short burst of speed was tailor made for it. Conversely, though, the days of better roads by the late teens with more sustained speeds was just something the boilers could not handle. The improved road system, then, finally helped complete the doom of the steamer.[464]

Moreover, steam car makers simply did not anticipate the long-range possibilities of consumers. In short, they were simply caught unawares. When an irate customer complained about his Stanley's small fuel tank, the Stanley brothers tartly informed him, "But, my dear Sir, three gallons will take you thirty miles, and that is all you will ever want to drive in one day."[465] Besides, steamers still had to burn some liquid fuel to make the steam. In fact, "they burned as much petroleum fuel as did the gasoline automobile."[466]

As for the electric cars, most of them stayed within the friendly confines of urban areas. Few people in rural America doubtless ever laid eyes on one, even at the beginning. It was even postulated, once upon a time, that electric and gasoline cars were not in direct competition. This weak attempt to compartmentalize the situation was both futile and short-lived. Said one confused contemporary source, the, "electric vehicle does not compete with the gasoline car for touring. . ."[467]

Some may, indeed, have felt *that* way, but the cold fact every gas car purchase took away one potential electric car sale. There is one final, often overlooked, mitigating factor against the early spread of electric cars into America's heartland. Much of rural America still did not have electricity. "Although Thomas Edison's inventions electrified cities in the 1870's, rural electricity was sparse until well into the 1920's, 30's[,] and 40's."[468]   For these rugged farmers, gasoline engines were probably *more* familiar to them as a group than electricity itself.

Sadly, once the electric cars started to fade into oblivion, their disappearance did not take long. In 1921, when a few electrics were *still* being made, "out of 9 million passenger cars registered in the United States, only 18,184 were electric."[469] Since steam car production never approached that of electrics in America, their percentage of total registrations declined even faster. Within a few years, the number of alternative cars registered in the United States had declined to the point where separate statistics were no longer kept. We are unable, however, to state decisively whether the early registration records were complete, since each state and/or municipality kept their own.

By the way, Detroit's ultimate victory of becoming the center of American car manufacture likely had more to do with the bankers and their often liberal lending methods and policies than anything else.[470] Ironically, car development money, which had been tight until about 1906, flowed more freely after that from the more open-minded Detroit financial officers. The records are filled with ambitious gas car pioneers seeking, and getting, money for developments aplenty. Most of the conservative New England and Eastern banking interests preferred to loan their monies to the steam/electric car concerns because they felt, as a group, the noisy, smelly, even repulsive gas buggy would be eventually driven from the market by their more "sophisticated" brethren. Time would more fully bear out this costly error in judgement.

In the final analysis, so soon as the alternative cars became of more trouble than they were worth, they were doomed. Not that there was anything basically wrong with them. But, the internal combustion car could do the same job with much less fuss. Its fuel was portable. It did not require the very frequent recharges like the electrics. These latter expended much of their energy carrying the heavy batteries. Nor did the internal combustion car require the long start-up times and frequent maintenance of the steamer. These last qualities gave the internal combustion car a degree of practicality not possible with the other types. Lastly, its motor was compact; leaving most of the rest of the car for other uses. We have already observed its high efficiency. Ultimately, the gas car won out because it deserved to.[471] For the technology of the time, there was no better choice. In

174

the end, it **did** become, "everyone's car."[472]   At least, most everyone's.[473] Finally, the timing of the Great Depression helped take away the alternative car's market.[474]   The people who often purchased the alternatives had other distractions in the early '30's. Like survival.

To wrap things up, long before the practical self-starter appeared, the i.c. car had decisively won the battle for supremacy among the three alternatives.

Footnotes

[1] Thomas LaMarre, "Sewing Up the Steam Car Market: White Steam," *Automobile Quarterly* 31, no. 2 (Winter 1993), 13-14.

[2] James J. Flink, *America Adopts the Automobile, 1895-1910* (Cambridge: MIT Press, 1970), 234.

[3] Sheldon R. Shacket, *The Complete Book of Electric Vehicles* (Chicago: Domus Books, 1981), 11.

[4] Michael Sedgwick, *Early Cars* (New York: Octopus Books, 1962), 21-22.

[5] E. T. Heald, *The Stark County Story* vol. 3 "Industry Comes of Age, 1901-1917." (Canton, OH: Stark County Historical Society, 1952), 3.

[6] Indeed, one of the first in the entire state.

[7] Heald, vol. 3, 281.

[8] Heald, vol. 3, 295.

[9] In fact, it muddied the waters further.

[10] Beverly Rae Kimes, *Standard Catalog of American Cars 1805-1942,* (Iola, Wisconsin: Krause Publications, Third Edition, 1996), 30, 78.

[11] *Cleveland City Directory, 1902,* 62. Was this the same Henry J. Aultman who moved to Canton in 1901 after constructing experimental steamers in Cincinnati? This is the assertion of *Standard Catalog,* in all three editions. Well, then, did he ever live in Cleveland?

[12] ibid., 40.

[13] The writer contacted the book's author, Beverly Rae Kimes, to provide amplification on this misunderstanding. Evidently two different cars emerged here. She promptly forwarded two items about the Altman construction, not to be confused with our Aultman Steam. In the first, an H. J. Altman, of 11 Pier Street in Cleveland, was listed among numerous other "experimenters" ("Trade News: Invention in Cleveland," *The Automobile & Motor Review* 7 (June 28, 1902), 15.) The second source displays a copy of the license certificate issued to the 32-year old mechanic for a, "Gasoline Touring Car." It contains a photo, and the engine was a two-cylinder gasoline unit built by Altman himself. (Larry Hawkins, "Those Were the Days!" *Cleveland Plain Dealer Pictorial Magazine* August 3, 1952, 12-14.) The existence of *this* particular individual and his car is thus established beyond doubt. In Cleveland.

[14] *Canton City Directory, 1901,* 109. One U. S. Grant Altman.

[15] The 1901 rendition details three Aultmans: the Aultman Company at 920 S. Market Street; the Aultman hospital; and Cornelius's widow. (ibid., 114.)

[16] *Canton City Directory, 1901,* 114.

[17] *Canton City Directory, 1902,* 125.

[18] Lorin Bixler, *Cornelius Aultman, C. Aultman & Co., and the Aultman Company* (Enola, Pennsylvania: Stemgas Company, 1967), 12.

[19] Richard Wager, *Golden Wheels: The Story of Automobiles Made in Cleveland and Northeastern Ohio, 1892-1932* (Cleveland, Ohio: Western Reserve Historical Society, First Edition, 1975).

[20] F. Scott Bailey and the Editors of Automobile Quarterly, *The American Car Since 1775,* (Princeton, New Jersey: Princeton

Publications, 1971), 241. The "Altman," which we have already looked at, is on page 235.

²¹ *Cleveland City Directory, 1905,* 43. No Henry J. Aultman is to be found anywhere in the 1905 city directory. The only Aultman listings are the Cleveland branches of both the Aultman-Taylor Machinery Co. and of the Aultman Engine & Thresher Company. Likewise, the name Henry J. Aultman does not appear in the 1905 Canton directory.

²² Kimes, *Standard Catalog,* Third Edition, 78.

²³ Heald, vol. 3, 295.

²⁴ *Dayton City Directory, 1900-1901,* 1096.

²⁵ Dave Emanuel, "Harry Clayton Stutz, 1876-1930: The Man, The Enigma, and The Legend," *Automobile Quarterly* 20, no. 1 (Fall 1982), 234-235.

²⁶ Heald, vol. 3, 295.

²⁷ We have yet another riddle here. According to *Standard Catalog,* "by September [1901,] he was experimenting with a small tiller-steered gasoline buggy." That "he" was evidently this same elusive Henry J. Aultman. According to a letter from Beverly Rae Kimes, the information came from the September 1901 edition of *Cycle & Automobile Trade Journal.* (Letter from Beverly Rae Kimes, September 28, 2001 to the writer). The writer searched that issue for any reference to an Aultman, of either incarnation. He found nothing. When the writer contacted Mrs. Kimes for elaboration by phone on October 24ᵗʰ, 2001, she could offer no further leads and could not recall specific page numbers. Not to take anything away from such historians.

²⁸ A check of Indianapolis City Directory for 1904 shows a Harry C. Stutz as residing at 1816 Shelby Street, his occupation a machinist. In the next year's directory, he was a salesman living at 2213 Bellefontaine; the only other Stutz listed in either year was

his cousin, Charles E. Stutz; he boarded off of Harry. (Raymond A. Katzell, ed., *The Splendid Stutz* (Indianapolis, Ind.: The Stutz Club, 1996), 8). As for this Theodore Reefsnyder, this writer has been unable to locate any information about him other than Heald's veiled, mysterious reference.

[29] Evidently, this gentleman was not the one affiliated with the law firm of Lynch, Day & Day, despite being so listed in one source (Heald, vol. 3, 295.) The 1901 city directory identifies Austin Lynch, of 1360 N. Cleveland Avenue, as the law partner. (*Canton City Directory, 1901, 343.*)

[30] Bixler, 57.

[31] Heald, vol. 3, 295.

[32] There are two copious files on the old company at the Stark County Historical Society. The files contain old sales catalogs, newspaper clippings of various events associated with the old firm, lists of employees, testimonials from various people who had purchased company products, along with a copy of Bixler's book. These appear fairly complete, except for any mention of the Aultman Steam. Surprisingly, not one word about it appeared in either of the two big files. A search of three hours was not productive, although Healds' original handwritten article from *The Stark County Story* was discovered in a vertical file in the back room.

[33] This is most unfortunate as there were a couple of persons known to be associated with the old company interviewed by Heald as he prepared his research materials. The cited article made mention of interviews Heald had conducted in his chapter's preparation.

[34] *Canton City Directory, 1898, 121.*

[35] *Canton City Directory, 1899, 118.* Apparently no 1900 directory was made.

[36] *Canton City Directory, 1901,* 114.

[37] ibid., 539.

[38] *Canton City Directory, 1902,* 603.

[39] A visit to the NAHC on February 3d, 2001 revealed little more information. The original of a photo of the car from *Standard Catalog* was found there. There are no brochures or repair manuals extant on this marque; assuming any were ever made.

[40] "The Aultman Steam Carriage," *The Horseless Age* 8 (September 25, 1901), 543.

[41] Kimes, *Standard Catalog,* Third Edition, 78.

[42] Herbert L. Towle "Aultman 4-Wheel drive Steam Trucks for Commercial Use," *The Automobile & Motor Review* 7 (November 1, 1902), 6-9. The experimentation work was done under the tutelage of a Mr. E. A. Wright, who probably designed the whole affair. The rear axle was driven by a pulley from the centrally placed motor; it seems fairly simple. For the front axle, though, Wright used bevel gearing attached directly to each wheel, while keeping the solid Ackerman system of steering knuckles. It was a rather novel idea in and of itself, without the four wheel drive. Of course, the forward mounted boiler and steam equipment balanced the whole package. Unfortunately, the one truck that was completed is no longer extant.

[43] *Horseless Age,* 543. The truck idea died aborning, despite the excellent features cited above. No doubt the continuing poor financial condition of the firm caused this circumstance.

[44] "Brief Notes of the Industry," *Motor Age* 3 (January 30, 1901), 943.

[45] An early type of fuel gauge.

[46] "The Aultman Steam Carriage," *Cycle & Automobile Trade Journal* 6 (June 1, 1902), 74.

[47] Mr. Karl Harsh, a 78-year old retiree at this writing, was the holder of so much trivial information about Canton's past he was christened "Mr. History" by Gary Pelger, then curator of the Canton Classic Car Museum.

[48] Mr. Harsh, however, could not recall any specific details about the Aultman firm and had never heard of the steamer. (Telephone interview of John Harsh, Karl's son, September 17, 2001.)

[49] "Aultman in Receiver's Hand," *Motor World* 9 (September 29, 1904), 11.

[50] *Canton City Directory, 1903,* 125.

[51] This writer has a copy of a form letter from the Stark County Historical Society file, dated simply "1903," sent out to 'Threshermen and Mill Men.' This was intended as a cover letter which the firm sent to prospective customers. The letterhead repeats the above listing, adds 'Saw Mills,' then the etc., but no automobiles.

[52] The Aultman's will and good faith, as it concerned the thresher operations, "was later bought out by Allis-Chalmers." (*Floyd Clymer's Album of Historical Steam Traction Engines and Threshing Equipment No. 1* (New York: Bonanza Books, 1959), 30).

[53] Bixler, 79.

[54] Interestingly, *The American Car Since 1775* lists only one Gramm in its "5000 Marques" list. This was an air-cooled cyclecar made by the Gramm Motor Truck Company of Walkersville, Ontario, Canada. This make was listed for 1913 only, with a twin-cylinder air-cooled gasoline engine. It bore no apparent relationship with the Ohio product. (291).

[55] *Chillicothe, Ohio: 1796-1996 Ohio's First Capital* (Jackson, Ohio: Jackson Publishing Co., 1995), 85.

[56] Ross County Historical Society, 45 W. Fifth Street Chillicothe, Ohio. "Population."

[57] Patricia Medert *Stories from Chillicothe's Past* (Chillicothe, Ohio: Privately Published, 1998), 120.

[58] This same business man would build the Gramm Steamer.

[59] *Eighth Annual Old Fashioned Day: Yoctangee Park September 3, 1978* (Chillicothe, Ohio: Craftsman Printing, Inc., 1978), 6.

[60] At that time, young Gramm was three years into a seventeen-year career at the First National Bank in Chillicothe. In the city directory of 1891-92, Gramm has two listings. One entry had him at 82 W. Fifth Street, with bookkeeper noted as his occupation. The other record specified his residence as 45 W. Fifth, this time as a messenger. (*Chillicothe City Directory, 1891-92,* 81.) If nothing else, this circumstance shows due caution must be used with the city directories. One thing was for sure, Gramm must have spent long hours working at the bank.

[61] "The Story of Benny Gramm... Pioneer Auto and Truck Builder," *Ohio Motor Travel* (Greenfield, Ohio: Automobile Clubs Publications) 28 January 1955, 4-15, inclusive.

[62] Hiram Percy Maxim, *Horseless Carriage Days* (Boston, MA: Little Brown, 1937), 1-2. And to likely thousands of other aspiring "inventors" as well. Maxim constructed some of the earliest of the 'horseless carriages.' He was familiar with all three types of power.

[63] Benjamin A. Gramm "Gramm Trucks... Trailers," *The Ohio State Journal* January 25, 1936, inclusive, no pagination. In fact, we again encounter controversy about key elements in the

story of the Gramm Steamer. Just like in the case of the Aultman Steam of Canton.

[64] In an interview for the *Scioto Gazette,* in May, 1938, Ben Gramm recalled his early days in the industry. He mentions that he built the first Logans in the new factory, apparently in 1900. However, Gramm's memory appears faulty on more than a few points.
For starters, the Logan was launched only about 1904; he may have meant his first full-scale foray into the business, i.e., his steam car outfit. Curiously, Gramm, in this same account, puts 1903 production of cars and trucks in the United States at, "1242." (*Scioto Gazette* May 7, 1938, Sesquicentennial Edition.) An example of misinformation at its best. The writer's own extensive research into this field puts car production for the same year at "approximately" 11,235. Or nearly ten times Gramm's figure.

[65] ibid. This was none other than Alexander Renick, listed as the President of the Mutual Loan & Savings Association and the First National Bank. (*Chillicothe City Directory,* 1900, 172). In fact, according to Gramm, "the final drawings for a six-passenger carry-all were finished by me at my desk. . . [and] Renick, then president of the bank, looked over these drawings." (*Scioto Gazette,* May 7, 1938.)

[66] Gramm, inclusive.

[67] The greater part of two days' worth of searching newspaper microfilm yielded no other information about the venture.

[68] *Eighth Annual,* 6.

[69] "Motor Storage & Repair Co.: The Pioneers in this Line of Work in Southern Ohio," *The Scioto Gazette* October 28, 1901. The business was not actually chartered until January, 1903, "under laws of South Dakota with B. A. Gramm and Joseph Schilder as incorporators." (John R. Grabb, *Little Known Tales of Old*

Chillicothe and Ross County, Ohio (Chillicothe, OH: John Webb Printer, 2001), 52.) Curiously, this particular book was printed in the same building as the Gramm Steamer was built, almost a century earlier.

[70] *Chillicothe City Directory, 1900,* 95. Williard A. Hall, who incidently practiced out of his home, was an outspoken advocate of the 'horseless carriage.' This was a trait he shared with many in his profession. It was, after all, a natural alliance. The coming of the car allowed rural physicians the ability to greatly enlarge the number of distant patients they could see in one day.

[71] Brown, *Ohio Motor,* 4. "Benny Gramm. . . uncrated the parts. He and Bob McVicker[s] spent the winter putting the machine together." (Martha Filipic, "Dr. Hall Owned First City Car in 1901" *The Chillicothe Gazette* October 10, 1985.)

[72] *Eighth Annual,* 6. As it worked out, Mr. Houser and Gramm would collaborate on quite a number of projects. A local historian indicated Mr. Houser constructed a steam engine around 1885, when he was just 15. This experience with steam propulsion undoubtedly influenced his friend Gramm to proceed with steam power for his first cars. In later years, Houser's widow donated this first steam engine to the Ross County Historical Society following her husband's death. (Interview of Mr. Randall Grabb, August 20, 2002.) Interestingly, Mr. Grabb (1915-   ) had not been aware Gramm had ever constructed a steam car.
Houser was not listed in the 1900 directory, but was in the *Chillicothe City Directory, 1904-1905,* on page 125. He resided at 150 N. Mulberry. From all indications, the early car pioneer was always mechanically inclined. He was also a key player in the Gramm Steamer enterprise. In addition to the Gramm, he would also lend his expertise to the Buckeye and the Logan cars. Later, in 1908, he became Chillicothe's first Ford dealer. Finally, Mr. Houser, "opened the town's first filling station in front of his machine shop and garage." (*Chillicothe, 1796-1996,* 277.)

[73] *Chillicothe City Directory, 1900,* 89.

[74] ibid., 183.

[75] This article appeared in the June 21, 1995 edition. The author of the article, Patricia Medert (1931-   ), did much original research on the activities of Ben Gramm, among many other important events in Chillicothe's history. At this writing, she worked at The David McCandless McKell Library division of the Ross County Historical Society. When this writer met Ms. Medert, she helpfully provided what information she had come across on Gramm's early automobile efforts, specifically the Buckeye and the Logan. Interestingly, she had not been aware of his building a steam car. This reinforces our earlier supposition about the obscurity of auto history. Most especially when two local historians had not found important information about a local notable. Not to take anything away from such historians.

[76] Moreover, as we have observed, this was not the extent of Gramm's car building activities. The much better-known Logan car, built from 1904 to 1908, was also Gramm's; its only extant example survived at this writing in the hands of the Ross County Historical Society. But this latter car had a small gasoline engine. Outside of the parameters for this thesis.

[77] Brown, *Ohio Motor,* 4.

[78] This image, found in Gramm's article in *The Ohio State Journal,* shows seven men, of which six were attired as workers, in front of the shop. Among those were Ben Gramm and his son, Willard. The only vehicle in view is the aforementioned truck. We can not tell what was going on *in* the shop, of course, from this view.

[79] Brown, *Ohio Motor,* 4.

[80] October 28, 1901, *Scioto Gazette,* 11. Some auto makers would merely take the best proprietary parts to assemble their own car; hence the term 'assembled car.'" The degree to

which proprietary parts were employed in the Gramm is not known.

[81] "Gramm Steam," *Cycle & Automobile Trade Journal* 6 (January 1, 1902), 168. Gramm was one of the very few auto makers who offered to custom build a power source for the customer. However, his earliest creations were **all** steam powered; with the possible exception of the suspect 'electric wagon,' just referred to. Moreover, according to local historian Floyd Brown, "Mr. Gramm always expressed a thought that the 'electric would come back some day.'" (Brown, *Ohio Motor,* 4.) This comment must mean Gramm had at least thought about making an electrical vehicle. This would have been even *after* the gasoline car became his preferred vehicle of manufacture.
This may also explain why the businessman did not specifically mention a steam car in his later recollections. In fact, he even failed to say how **any** of his cars from 1902 were powered.

[82] A glimpse at the two contemporary articles betrays some important clues along this line of reasoning. For instance, both use similar terminology. The *Cycle* piece states of Gramm & Company, "they put the most painstaking and conscientious work on their vehicle. . . [feeling] they have the nearest thing to perfection." (*Cycle & Automobile,* 168.) In the *Scioto Gazette,* they will be, "putting up. . . one of the best machines which is now on the market." (*Scioto Gazette.*) The two quotes appear to emanate from the same source, likely Gramm himself.

[83] With financial help and some friendly "advice" from a wealthy local businessman, a certain George Hunter Smith, Gramm found the necessary funds *and* a ready venue.

[84] Both of these gentlemen were well-known in the local area. Nonetheless, the writer in *The Scioto Gazette* was being a little euphemistic when he referred to them as, "men who are artists in the different mechanical lines pertaining to this fast-rising industry." Mr. Robert McVickers (1879-1947), who resided at 93 S. Walnut—within sight of the Gramm factory—was actually

the manager of The Duncan Steam Laundry. His 'day job' undoubtedly meant he could devote less time to the automobile than he would have liked. However, the plant's proximity to his home no doubt allowed him some latitude in this regard. (*Chillicothe City Directory, 1904-1905,* 157.) Mr. Frank Ramsdail, also of 93 S. Walnut, was listed as a 'bodymaker' in the same year's city directory (ibid., 188.) Mr. McVickers boarded off of Ramsdail. All of our known sources and the few photos we have of these individuals strongly suggests they had a close working relationship. Collectively, the four businessmen constituted the best automotive talent in Chillicothe at the time. However, no further specific details have been discovered. Moreover, there were still other locals involved in this company. These included Mr. Harry Hermann, the firm's "first bookkeeper," and Otis Story, duties unspecified. (*Scioto Gazette,* May 7, 1938.)

[85] *Ross County Recorder* "The Story of the Logan Automobile," October 2000 (Chillicothe, Ohio: Ross County Historical Society, 2000), 3. Incidently, the Buckeye was also made for just one season, 1903. (Kimes, *Standard Catalog,* Third Edition, 157.)

[86] This was the carryall we have already alluded to. The Buckeye was an eight-passenger style. Such types would not have resembled each other.

[87] Just how many of these cars were Gramm Steamers we are not precisely informed, but undoubtedly they are the same cars the reporter from *Cycle & Auto Trade Journal* scrutinized. In line with this, and making a reservation for Gramm's sketchy memory, we can take the liberty of adjusting the production date of Gramm's cars to around October of 1901. Two of these cars were exhibited at the Madison Square Garden Auto Show in January, 1903, according to the article. Once more, though, we have problems. A check of the *1903 New York Automobile Show Booklet* (New York: Association of Licensed Automobile Manufacturers)

failed to list a Gramm, a Gramm-Schilder, even Gramm-Smith, or any form thereof as an exhibitor. In addition, no firm specifies Chillicothe, Ohio, as its base. This is another of those many mysteries we have already looked over. The term "motor cars" was used here; no mention of trucks/commercial vehicles. However, we can not say this for sure either way. Of course it is always possible Gramm's exhibition got in as a last-minute addition. However, this is nothing more than just a probability. Nonetheless, Gramm specifically remembered conversing with tire magnate Harvey Firestone at the show.

[88] Kimes, *Standard Catalog*, Third Edition, 651.

[89] *The Scioto Gazette*, October 28, 1901, 11.

[90] *Cycle & Automobile*, 168.

[91] No information was extant on the operating range.

[92] *Scioto Gazette*, 11.

[93] At this writing, the structure housed Webb's Print Shop to be exact. The writer took in the view from the vantage point of the photos. With these as a reference, we can see the building had little changed; except for some reworking of the wood trim around the doorway. Even the alley behind the shop had not fundamentally changed. Despite the passage of nearly a century. One could almost picture Gramm and his workers laboring away on their cars in those far off days.
When this writer visited the location on February 28th, 2001, he inquired about the building's history. The present building owners knew a car factory had once been operational there, but little else. The writer was told of some leftover mascots and business paraphernalia from those days on the building's third floor, but he could not see them since the tenant said the upper levels were, "a hazard to walk around on." (Writer's notes.) The first floor had no apparent traces of what had once been there.

[94] And he unquestionably helped shape the American motor truck industry.

[95] See page 49.

[96] He was as proud of this fact as he was of anything he ever did automotive. (A. S. Krotz, "Reminiscences of An Old Timer," *Old Timer's News* 3 (July, 1945), 28.)

[97] And he also happened to be one of the better educated inventors of his day. Krotz graduated from Valparaiso University before he set out to seek his fortune.

[98] Advertisement, Willard Storage Battery Company, found at NAHC. Krotz stated, "During 1901 I completed an electric automobile for myself and one for. . . the Willard Storage Battery Company, Cleveland, Ohio." ("Reminiscences," 28.) These two vehicles were to be the first of several constructed for the Willard firm.

[99] Advertisement, Willard Storage. Obviously, this reference was to the battery's range; i.e., the interval between charges. Already we can discern the first awareness of competition in the car building business. Even if the caption here referred specifically to the battery itself.

[100] "Reminiscences," 28-30.

[101] "Alvaro S. Krotz," *A Biographical Record of Clark County, Ohio* (New York: The S. J. Clarke Publishing Company, 1902), 324. Krotz said this was a toy that, "moved along at about a slow walking speed." ("Reminiscences," 28.)

[102] *Biographical Record,* 324.

[103] Almost simultaneously Alvaro filed for a patent on, "a steering and power control." (Beverly Rae Kimes, "Behold the High Wheeler, The Sears for Instance," *Automobile Quarterly* 9, no. 1 (Fall 1970), 112.) This was not an isolated event. Krotz,

"concentrate[d]. . . on the improvements of wheels, axles, [and] bearings." (Don McCray, "Alvaro S. Krotz and the Sears Motor Buggy," *Antique Automobile* 20 (Fall 1956), 27-28.)

[104] The latter firm thrived on building early buggy outfits under contract for other companies.

[105] "Reminiscences," 28.

[106] Don McCray, "Biography of a Pioneer," *Goggles & Gaunlets* 4 (September 1956), 2-4.

[107] And this was not just from the citizens of Springfield, either. Krotz's experimenting certainly did not take place without widespread publicity. As the contemporary entry stated, "Mr. Krotz's knowledge of the great mechanical principles. . . and his work have called to him the attention of the business world." (*A Biographical Record of Clark County*, 325.) The wily inventor worked hard to capitalize on his new found renown.

[108] *A Biographical Record of Clark County*, 325. The aforementioned ad would place the date of manufacture of Krotz' little machine as approximately late 1899. McCray states there were a couple of electric automobiles built, "which incorporated the devices illustrated in this [first] patent. . .." (McCray, "Sears Motor Buggy," 27.) He had a number of automobiles already to his credit, as we have seen. Perhaps those electric cars were part of the Willard order. We do not know if there were two different models, or just one type. In fact, Krotz had already sold 12 electric cars to the Willard company around 1901-1902. (Information supplied by John R. Link, of Indiana.) Mr. Link is a relative of the late Mr. Krotz—the latter was his great uncle. (Interview of Mr. John Link, October 23, 2003). Willard reportedly used some of these cars to, "experiment with cord versus fabric tires." (McCray, "Sears Motor Buggy," 27). This firm was able to prove, beyond doubt, the superior design of the cord tire using the cars built by Krotz. Some of the machines must have been used up in the process. The rest

were sold to private owners. Further elaboration on this topic is not available.

[109] *Springfield City Directory, 1902, 320.* There is nothing else that can be gleaned from this particular year's directory.

[110] "In March of 1903[,] Alvaro Krotz announced. . . he was building two electric cars." This has to refer to private sales directly to consumers. (Kimes, *Standard Catalog,* Third Edition, 831.) Evidently these sales were done without benefit of a dealer network.

[111] Bailey, *The American Car,* 308. Significantly, Larry Freeman's (1904-1995) *The Merry Old Mobiles* does not carry a listing at all for either form of this marque in his roll of American cars built between 1895 and 1950. (Watkins Glen, New York: Century House, 1949), 226-233.)

[112] This was likely called the Willard Storage Battery because that was the source of the development money, although this is not spelled out in any literature.

[113] Unfortunately, we do not know what the **actual** range of the battery was. We should also note that exaggeration in advertising was not a late twentieth-century invention.

[114] *Springfield City Directory, 1903, 342.* At this writing, there were a number of abandoned warehouses, old factory buildings in the immediate area. One of those edifices undoubtedly housed the enterprise, although precisely which one appears lost to history.

[115] ibid., 644.

[116] "News and Trade Miscellany," *The Automobile* 8 (June 27, 1903), 671. This was already underway, as we have observed. Magazine lead time likely played a key role here.

[117] ibid. Few details have come down about the design and other features of this battery, although it appears to have been

large. This was in common with most electric cars of the early days.

¹¹⁸ *Springfield City Directory, 1903,* 342. This rather generic term could encompass a wide range of mechanical products/services. Clearly, not just automobiles.

¹¹⁹ In point of fact, this particular vehicle rather more resembled a horse-and-buggy outfit without the horse. Much more so than the Aultman Steam and the Thresher Electric of similar vintage. Evidently this design was not satisfactory for Krotz; for the later design which he offered to Sears looked less like a buckboard. The center of gravity on the earlier Krotz electric was much higher; the passengers therefore were further from the ground. The battery storage compartment of the earlier electric was, of course, also functional. It was far more prominent than the rear design of the Sears buggy. The latter car did not require an extra large battery compartment.

¹²⁰ The driver and one passenger. There was no provision for rear seating.

¹²¹ "Early Old Timer Still Believer in Cheaper Cars," *Old Timer's News* 3 (August 1948), 10. Alas, there are no reports extant on the reaction of any who literally tried to take a bite out of the notched tire. In fact, it makes one wonder what such "curious" individuals hoped to accomplish from such a deed.

¹²² The old horse-drawn buckboard had never required such advanced tire designs. Increasing speed capability associated with the automobile had created this need.

¹²³ *Automobile Review* specifically mentioned the Krotz firm was, "manufacturing a solid cushion tire for the automobile which has some special features." (*Automobile Review,* 4 (October 15, 1903), 163.) Unfortunately, this source does not then state just what the special features were, although we can safely assume it was referring to the tread itself. (At that stage, as we have seen,

tires had no tread on them; they were smooth.) However, this piece does mention the tire was being used for a new car of Krotz's own manufacture.

[124] If he ever seriously considered a steamer, the writer has not been able to find any evidence of his intentions thereof. In fact, Krotz, "built a gasoline car for a doctor [in 1902]." (Link Interview, October 23, 2003).

[125] *Springfield City Directory, 1905,* 364.

[126] It is not known what kind of car manufacture, if any, took place at this new location. In the 1906 directory, Krotz Manufacturing still warranted a listing (372), but in the business section, there was only automobile manufacturer listed, the Springfield company (700). (*Springfield City Directory, 1906.*)

[127] The inventor would claim, years after the fact, he built, between 1907-1908, a total of 13 of these cars. The vehicles themselves did not have nameplates. Apparently these cars were prototypes. (Don McCray, "Biography," 2.) It would seem at least a few of them were sold by Sears; they were built at the factory later used for all the Sears' cars.

[128] As an interesting aside, when the new car was first shown to the public in showrooms, the Sears & Roebuck Co. quickly realized they could do more business if they set up retail outlets. These branches, originally founded for the Sears High Wheeler, long outlasted the car. There was a diverse grouping of merchandise on display. In fact, many years later, after the mail order business was phased out, Sears banked its entire survival on the retail store business.

[129] In fact, the Sears enterprise was a perpetual money loser throughout its existence. Production ended in 1912.

[130] "Auto Factory for Defiance," *Daily Crescent News* February 4, 1908.

[131] The one thing the manufacture of cars required, even in those early days, was a healthy dose of finances. The gentlemen Krotz enlisted in this latest endeavor all appear to have had the necessary wherewithal. *The Defiance City Directory of 1909* was invaluable here. For instance, Mr. John A. Deindoerfer, publisher of the city newspaper (76), and John C. Diehl, of the Christ Diehl Brewing Company (79), were of this ilk. Local banking interests existed in the person of C. Perry Harley, president of the Merchant's National Bank, in 1909 (100).

[132] Two of which were stillborn.

[133] Kimes, *Standard Catalog,* Third Edition, 703.

[134] ibid., 424.

[135] "Minor Mention," *The Horseless Age,* 14 (February 2, 1908), 189.

[136] "To Make Buggies," *Motor Age* 9 (February 20, 1908), 29. In fact, both trade publications mention a starting capital of $50,000. Obviously, more would be required to make a serious effort at manufacture. The local newspaper was specific about Krotz making full use of previous inventions for which, "he is now receiving large royalties." (*Daily Crescent,* February 1908.)

[137] *Daily Crescent,* February 1908.

[138] "Buggy Company Formed at Defiance, O." *Automobile Topics* 16 (February 22, 1908), 1516.

[139] *Daily Crescent,* February 1908.

[140] *The Defiance City Directory, 1907-1908* listed this man as Secretary and Treasurer of the Defiance Grocery Company (49). Others in town of the name of Krotz included a Charles W. Krotz and Alva Krotz. These two were listed as partners in real estate. (ibid). The existence of such a diverse group of investors, both

relatives and others, ensured the new car would be accorded its share of attention.

[141] This individual was the great grandfather of John Link. The writer interviewed the latter in preparation for this work. Mr. Link provided much insight on the era.

[142] *The Daily Crescent News,* in a general listing of past local businesses, noted the location of the factory and identified its production as, "engine driven buggies." (*Daily Crescent News* October 22, 1938.)

[143] *The Horseless Age,* 189.

[144] At this writing, there was no trace left of the automobile plant, which was demolished years ago. A check of the Sanborn Insurance Maps (provided by the courtesy of the staff of the Defiance Public Library) showed a "skating rink" at the sight in 1905. In 1911, "there is an 'electric motor' business." (Source: Defiance Public Library.) A professional building and a paved parking lot were the occupiers of the former factory location at this writing. At the time Krotz's company was there, the street was named Front Street. At this writing, it was known as Fort Street.

[145] *Daily Crescent,* February 1908.

[146] ibid.

[147] *The Daily Crescent News,* October 22, 1938.

[148] "A. S. Krotz Among 36 Men Honored by Auto Industry," *The Rockton Herald,* May 16, 1946. By 1946, Krotz had moved to Rockton, Illinois.

[149] At this writing, Route 534 substantially flowed between Witmer's Feed & Seed and the Garfield Slag Company. In 1949, an elaborate train bridge was built, according to the then current owner of the slag company, Richard Saulino. (Interview of Richard Saulino, April 2, 2002.) At the same time, the road's course had

been shifted. The original Route 534 as it existed when the car was being built in Garfield ended, at this writing, at Witmer's. The remains of the old factory were about a mile from that business off the road. At this writing, the only way to reach the place was by foot or ATV (All-Terrain Vehicle).

[150] MacDonald Prospectus. No date.

[151] The two spelling variations, for some reason, are used interchangeably in the company literature.

[152] This firm "constructed" steamers between 1919 and 1922.

[153] "Gearless Car to be Made," *Automobile Trade Journal* 24, Part IV. (June 1920), 223.

[154] We have detailed bare chassis illustrations for both the Gearless and the MacDonald. There are a number of important distinctions between the two. For instance, the Gearless utilized two 2-cylinder engines, one connected to each back wheel (guess they were not kidding about that 'gearless' stuff, eh?) By comparison, the MacDonald was the more conventional; if that term can be applied to steam cars in the early twenties. (Edward J. Blend, Jr. "The Gearless: The Steamer was a Swindle," *Antique Automobile* 34, no. 5 (September-October, 1970), 41.)

[155] There were not very many steamers at all on the market by then. Floyd Clymer, for example, could concentrate most of the available models from 1923 on a single page of one of his scrapbooks. ("A Quintette of Steamers," *Floyd Clymer's Historical Motor Scrapbook,* Vol. 4 (Los Angeles: Clymer Publications, 1952), 171.) The illustrations included a Coats Steamer, a Detroit Steamer, a Stanley, an American Steamer from Chicago and, of course, a MacDonald Steam Bob-Cat.
This sudden reappearance of new, contemporary steam car marques in the late teens–early twenties (just when they seemed

down for the count) was a most curious phenomenon. When the writer spoke to author Keith Marvin, he inquired directly about this issue. Mr. Marvin was very pointed in his comments: "There is still a certain romantic charm about a steamer. People invested, bought stock in them [sic] because they felt steamers would eventually return someday. Even today, there is no good reason why steamers are not made. Unfortunately, most of those [early twenties'] efforts by steam car companies was to sell stock. That was the extent of it." (Telephone interview of Keith Marvin, August 28, 2001.)

[156] John Bentley, *Oldtime Steam Cars* (Greenwich, Conn.: Arco Publishing Co., 1953), 92. While we are on the subject, the MacDonald prospectus detailed financing for the firm. "Authorized Capital was $500,000." This was broken down by, "40,000 shares of Preferred Stock and 10,000 shares of Common Stock, at $10 per." (MacDonald Prospectus.) Unfortunately, no information surfaced on how much was subscribed or by whom. Nonetheless, it seems a fair assumption that most of the new company's cash still came from the "proceeds"of the Gearless enterprise, in spite of the new stock issuance.
It might also be worth noting, in those days before the Securities and Exchange Commission, there was a certain amount of leeway tolerated between "inventors" and potential investors that was not possible later. At best, the results were generally less than satisfying. At worst, the combination of a still often gullible public and unscrupulous promoters developed a number of, say, "headlines" in those more innocuous times. (Blend, 39.)

[157] *Auto Trade,* 223. These cars were likely all Gearless models, though.

[158] Keith Marvin, "The American Steam Car of the 1920s— A Reflection of a Time that was," *Antique Automobile* 52, no. 5 (September-October, 1988), 37.

[159] Keith Marvin with Arthur Lee Holman, *The Cars of 1923* (Troy, NY: Automobilists of the Upper Hudson Valley, 1957), 75.

[160] Mr. Lewton attended school in Salem with the daughter of Duncan MacDonald, and was an old family friend. He indicated they lived at 1715 State Street in Salem at the time he knew them. Mr. McDonald often made reference to the old steamers. A check of the contemporary Salem directory produced the first local address for Duncan MacDonald and his wife, Margaret, to be at 1416 McKinley Avenue in Salem. From all indications, MacDonald's sole purpose in moving into the area was to build his steamer. (*Salem City Directory, 1923-1924,* 242.) A former Kaiser-Frazer-Jeep dealer for Lisbon, William Lewton owned Lewton's Garage. At this writing, he had a 1923 Model 740 Stanley Steamer sedan under a canopy on the floor of his still operational garage. The writer owes Mr. Lewton a debt of thanks for supplying copies of a packet of rare and otherwise inaccessible materials he had compiled over the years on this obscure marque. Sadly, MacDonald's daughter, Alyse MacDonald Boehm, had died a short time before the writer's visit.

[161] "So designed as to overcome the troubles experienced in fire tube," read the Specifications section. A wise precaution, but the resulting influence was undoubtedly negative in weight.

[162] This is the only known sales literature ever released by the firm in Garfield; a simple four page printed flyer, done either by or for the company. The involved entity identified itself by the rather formal sounding title of Steam Vehicle Distributors, Inc., on the front cover. The car shown in the brochure was a, "MacDonald Custom Built Steamer." The displayed vehicle was a sedan, with four men (including Duncan MacDonald himself) looking admiringly under the hood at the imposing 20 X 16 in. Fire Tube boiler. The opposite page details a whole list of advantages of steam cars, much like the Stanley's latter-day ads. It also boasts, again rather immodestly, that the MacDonald car was, "the smoothest and most powerful Vehicle for its weight in America."

Surprisingly, the brochure then fails to mention the weight of the sedan in the brochure.

Unfortunately, the writer was unable to locate any specific information about MacDonald vehicle weight. However, the Gearless car was reported weighing 2850 pounds on a 126- or 127-inch wheelbase. The MacDonald car, being close in appearance although of slightly different construction, likely boasted a similar weight. (G. Marshall Naul, *The Specification Book for U. S. Cars 1920-1929* (Osceola, Wisconsin: Motorbooks International, 1978), 124-125.)

Finally, the front page bears the question, "Do You Want a Custom Built Steamer?" How many people would want to answer that question in the affirmative by the early twenties must have been quite low.

[163] Bentley, 112. MacDonald called this arrangement, "patented direct drive." This was probably done for more balanced weight distribution forward and aft.

[164] The Steam Bob-Cat, as announced, was a 1923 model. A later prospectus exists, this one showing the above cited sedan, bearing a 1924 date. This was a penciled in reference, however. Interestingly, the "later" prospectus does not bear a printed date, either. Also, this photo is of the same model, same view as appears in *Standard Catalog*. (Kimes, *Standard Catalog*, Third Edition, 913). *Standard Catalog* cites the source of the latter photo as Keith Marvin. When the writer interviewed Mr. Marvin, the latter recalled an old advertisement. (Telephone interview of Keith Marvin, August 31, 2001). After some investigating among possible sources, the photo surfaced in *Motor*, in the January, 1924 issue (368). It appeared under the subheading, 'District Distributors Wanted for a Real Steam Car.' The first line of the ad states that, "A real opportunity awaits live district distributors. . .." (Now one would think being alive would be a requirement for selling cars!) This, however, does differ enough from the Gearless images to say it was not just a reworked Gearless. Moreover, this also does prove conclusively at least one sedan was constructed.

[165] Mr. Lewton recalled this particular problem the writer has not encountered anywhere else. It seems the rear-mounted four cylinder engine was hand-made, apparently at Garfield. There was a design flaw in which the center bearing between the two banks of twin cylinders prematurely wore out. This was a crankshaft defect. Unquestionably, more development time/money would have rectified the situation. So this hurdle was not a fatal one. Before the problem could be addressed, though, Mr. Lewton said the firm had already closed. (Interview of Mr. Lewton, August 2, 2002.) As for engine specs, the four cylinders were of 3 in. bore X 4 in. stroke, for a total of 113 cubic inches. No hp figures were ever released.

[166] MacDonald Steam "Bob-Cat" Prospectus. The reference to heavy steam cars was something; the Steam Bob-Cat car tipped the scales at a hefty 3100 pounds. (Bentley, 112.)

[167] This drawing, a rough sketch of the Steam Bob-Cat, is mostly drawn in with pencil, significant parts being duly labeled. One of the noteworthy features, evident even in the earliest rendition, and retained throughout, was the exhaust steam being returned to the condenser. By the twenties, all steamers utilized such devices to conserve water and reduce the number of refills.

[168] Naul, *Specification Book 1920-1929*, 178. But, "efforts to locate an actual photo of the [Steam Bob-Cat] machine have been futile." (Marvin, *Cars of 1923*, 75.)

[169] MacDonald Prospectus.

[170] Blend, 39.

[171] MacDonald Prospectus.

[172] "Demonstrations Show Steam Car a Success," *The Salem News* November 14, 1921.

[173] "McDonald Soon to Start," *The Automobile* 48, Part III. (May 10, 1923), 1050.

[174] Advertisement, *Motor* January 1923, 282. This same ad mentioned patents that had apparently been issued to Duncan MacDonald covering certain aspects of the MacDonald car construction like the, "Patented Direct Drive Engine-Rear Axle with [integral] Boiler-Burner unit[s]" that were intended for small commercial vehicles while the heavier units, "are ideal for converting gasoline-propelled Motor Busses, Delivery Trucks[,] and Pleasure Cars into Steamers." These unnumbered patents were dated September 12th, 1922, October 10th, 1922, and November 7th, 1922, according to the ad. What was covered by any or all of these patents is simply not known. Unfortunately, an exhaustive search of databases and patent indexes produced nothing. This certainly casts a dark shadow over MacDonald's colorful career.

[175] ibid.

[176] Interview of Mr. William Lewton, August 2, 2002. If these gentlemen ever truly felt a steam car, especially *their* steamer, would replace the internal combustion car, they were doomed to disappointment. In addition, we do not know the extent of the difficulties involved in preparing the car for the market. Unfortunately, the everyday workings at the factory are largely unknown. Nonetheless, a few tidbits from those days have survived to add spice to our account.
One story Mr. Lewton obviously relished was the time the throttle broke on the workmen while they were demonstrating their car to a potential customer. (The unfortunates may have forgotten the need for brakes on the car in this instance!) The workers had to run the car into a nearby field, turning circles until their machine stopped. On another instance, Mr. Lewton recalled the workers had to wait two hours at Youngstown to get steam up to return home. Apparently, the employees had traveled there to display the car.

[177] *Motor* January 1923, 282.

[178] Mr. Mather's garage was at one time the local post office for Garfield, and he was the postmaster. He was very much a local

citizen; in fact, he was born in a house across the street from where he resided at this writing.

[179] Interview of Mr. Earl Mather, April 2, 2002.

[180] He spoke of the whole incident with a sense of relish that was most refreshing, even surprising.

[181] *The Damascus Herald,* 3. When the writer spoke to Mr. Meiter's grandson about possibly interviewing Mr. William Meiter, he learned the elder Mr. Meiter had died many years before. (Telephone interview of Mr. Mark Meiter, August 14, 2002.)

[182] The writer has been able to locate only two sources putting production that high. The Damascus Area Historical Society, based in Damascus, Ohio, through its house organ, *The Damascus Herald,* had a write-up about the car. The article reports, "The Antique Automobile Reference Collection. . . at Philadelphia, PA, reports that approximately 48 McDonald autos were sold." ("Museum Acquires McDonald Steamer Display," *The Damascus Herald* 7 (Autumn 1992), 3.) When the writer checked with Mr. Stuart McDougal at the Philadelphia Free Library to confirm this statement, the latter e-mailed that he had nothing in the collection on the car and, as far as could be told, just one prototype was built at Garfield. He also indicated he had nothing to do with the Philadelphia reference and could not indicate its source. (E-mail, S. McDougal to H. Redman, August 14, 2002.) Curiously, *The American Car Since 1775* also reports, "About 48 cars were built." (Bailey, *The American Car,* 314.) Apparently, both reports stemmed from the same original source; but the writer could not locate or even identify it.

[183] Kimes, *Standard Catalog,* Third Edition, 913. The aforementioned *Motor* ad even said the firm's plans, "call for the manufacture of 1000 jobs." (*Motor,* (January 1924), 368).

[184] Naul, 178-179. Such kits were theoretically available to convert an internal combustion engine into a "steam" engine.

[185] Kimes, *Standard Catalog*, Third Edition, 913. In support of this contention, *The Automobile*'s reporter offered, "The first output of the plant will be [steam engines]. . . to replace gasoline engines in existing cars." (*The Automobile*, 1050.)

[186] Mather interview, April 2, 2002. A local resident, Larry Wallace, had purchased the sight of the factory long before, along with the surrounding land. As for the road leading to the abandoned sight, nature had long since reclaimed it.

[187] ibid. As a postscript to Mr. Mather's experiences with MacDonald, he said his brother Raymond was a welder in the local area once upon a time. Mr. MacDonald one day happened into his shop with a steel 'contraption' that closely resembled a maze. The gentleman was acting very evasive about the whole matter. Later, Raymond told Earl he knew what Duncan MacDonald was working on. He was experimenting with an automatic transmission of some sort. Sadly, the would-be inventor lacked the financial resources to fully develop the device. This was well before the first commercially successful modern automatic transmission. In fact that, "had to be still in the Twenties." (Interview of Earl Mather.)

[188] However, this was in connection with the Gearless proceedings in Pennsylvania. Not on anything going on in Garfield. Besides, just how much involvement MacDonald had with the various and sundry details of the Gearless is not known.

[189] The *Motor* ad from January, 1924 was the last for the company. As a point of some irony, those interested persons who wanted a MacDonald Steamer franchise should, "address your communication to the President [of the firm]." McDonald was behind bars by then.

[190] Interview of Mr. Lewton, August 2, 2002. The 1923-1924 city directory had no listing for this firm. However, in the *Salem City Directory, 1927-1928,* a Grate-Overland Company

owned by E. L. Grate was listed. The address was at 155 Garfield Avenue. (143.)

[191] Telephone interview of Mr. Stanley, August 9, 2002.

[192] Interview of Mr. Lewton, August 2, 2002.

[193] Bentley, 112.

[194] *Floyd Clymer's Historical Steam Car Scrapbook* (Los Angeles: Clymer Publications, 1945), 180-181.

[195] This would also explain why no Gearless officials made the trip to rural Ohio with Mr. MacDonald.

[196] The population of the city in 1900 was 85,333. Dayton & Montgomery County Public Library, 215 E. Third Street Dayton, Ohio.

[197] The other was the firm of Thresher & Co., a varnish manufacturer located at 863 E. Monument Avenue. *(Dayton City Directory, 1900-1901, 1120).* This firm was headed up by an Albert Thresher, of 41 N. Perry Street. (*Dayton City Directory, 1893, 752*). It never had any direct connection to the Thresher Electric car. We only mention the latter to show it had no relationship to the Thresher Electric.

[198] *Dayton City Directory, 1893, 752.*

[199] *Dayton City Directory, 1900-1901, 1120.* Finding the location of the Callahan Power Block itself took some sleuthing. While that entity is mentioned, there is no such spot identifiable from the directories. The book, *For the Love of Dayton: Life in the Miami Valley, 1796-2001,* Edited by Ron Rollins (Dayton: Dayton Daily News, 2002) was of assistance here. A photograph from 1890 shows East Third Street looking on from Main Street (page 80). At that particular sight, in 1892, the Callahan Bank Building was constructed. (Source: Dayton & Montgomery County Public Library.) The Callahan Power Plant, in the rear of

the Third National Bank of 21 East Main Street, appears as Print #315 in the *Lutzenberger Picture File, 0300-0449,* located at the main library. This was the apparent location of the Callahan Power Block. At this writing, a banking institution still occupied the sight.

As for the disposition of the Thresher Electric Company, it is listed as late as the 1905-1906 city directory, on page 1346. But, in the 1907-1908 edition, the concern is no longer found. We can safely surmise it must have ceased business somewhere in there.

[200] *Dayton City Directory, 1899-1900,* 1063.

[201] ibid., 1210. As for automobile manufacturing, that was yet to be accorded a separate heading in Dayton city directories.

[202] "The Thresher Electric Vehicle," *The Horseless Age* 6 (April 11, 1900), 14.

[203] "The Thresher Electric Company's Product," *The Motor Vehicle Review* 5 (April 17, 1900), 21.

[204] Unlike the other turn-of-the-century machines we have surveyed here, there exist a couple of excellent photos of the Thresher car, from a couple of good angles. They all confirm a first impression of toughness and durability.

[205] *Horseless Age,* 14.

[206] *Dayton City Directory, 1899-1900,* 1168.

[207] *Motor Vehicle Review,* 21.

[208] Moreover, as we have already observed, many of the early electrics resembled horse-and-buggy outfits *sans* the horse.

[209] *Horseless Age,* 14.

[210] Of course, there were other outfits interested in such features. "The importance of the principle of three point

suspension. . . has long been realized and many attempts have been made to employ it." (Hugh Dolnar, "The Marmon 1906 Touring Car," *Cycle & Automobile Trade Journal* 9 (October 1905), 116.) The Thresher's was yet another version of this type. By the way, independent front suspension really is just, "a sub-frame suspended from the main chassis frame." (Donald Clark, ed., *Anatomy of the Automobile: How It's Built—What Makes It Run* (New York: Galahad Books, 1976), 90-91. In the Thresher, this was obvious.

[211] *Motor Vehicle Review,* 21.

[212] In fact, the then contemporary term "road" was often just a euphemism for what was, at best, a cow path or trail.

[213] Robert F. Scott, "Does Mourning Become the Electric?," *Automobile Quarterly* 5, no. 4 (Summer 1966), 203. In this construction, the entire wheel/axle assembly revolved. The Thresher, on the other hand, featured a solid rear axle, each side being geared independently.

[214] *Horseless Age,* 15. The dual power plants are clearly visible in front of the rear wheels, and drive was implemented by an inner hub assembly to the driving wheels. Above this whole assembly, a side vent grille, housing the battery, was prominently visible. The passengers and the driver sat over this enclosed battery compartment.

[215] ibid.

[216] *Motor Vehicle Review,* 22. The speeds were three, seven, ten, and 15 miles per hour, which no doubt involved cutting off power from one or both banks of cells.

[217] Of course, such a "controller" was not unique to the Thresher Electric car. However, whether the method of control in this particular case involved power applied from battery cells in series or series-parallel was not discovered.

[218] Kimes, *Standard Catalog,* Third Edition, 1469. The car was likely a casualty of the shaky financial condition at the parent company.

[219] Morris McNeal Musselman, *Get a Horse!—The Story of the Automobile in America* (New York: Lippincott Company, 1950), 40.

[220] A then contemporary term for early pioneer enthusiasts.

[221] Carl Burgess Glasscock (1884-1942), one of the early auto historians, from the perspective of near contemporary times, wrote, "Few and inadequate records were kept. . . and no individual sufficiently interested to collate claims. . . [i.e., of the pioneers] until years later." (C. B. Glasscock, *Motor History of America* (Los Angeles: Clymer Publications, 1946), 17.))

[222] Ralph Stein, *The American Automobile* (New York: Random House, 1975), 173. Obviously, under these circumstances, little thought was given to recording events. Moreover, this kind of attitude appeared to be the norm in those days. By the way, not quite two dozen of the early Stearns' cars were steamers, before a total switch to gas cars occurred around 1900. (Bailey, et al., *The American Car,* 113.)

[223] T. R. Nicholson, *Passenger Cars 1863-1904* (New York: MacMillian Company, 1970), 4.

[224] "A Plea for Discussion," *The Horseless Age* 5 (March 28, 1900), 9. At first glance, the journalists appeared to be wholly neutral here. However, previous to this, the same magazine, through the auspices of its editor, E. B. Ingersoll, had been highly critical of the electrics. In particular, of what it called the 'lead cab trust' (a hazy reference to the Electric Vehicle Company's urban cab operations). "Storage Battery Financiering," *The Horseless Age* 5 (October 18, 1899), 6-7; James F. Bellamy *Cars Made in Upstate New York* (Red Creek, New York: Squire Hill Publishing Company, 1989), 159. Among other things, this article spoke of

the, 'heavy, inefficient, delicate[,] and destructible apparatus wholly unfit for general locomotion.' The veiled reference was to the lead battery, the heart of the electric car. We will shortly deal with the Electric Vehicle Company in a little greater detail. Notably, the article concluded with a prophetic statement: "The outlook for the lead cab is utterly hopeless."

So the editor of *The Horseless Age* had already taken a stand against the electric car and its heavy, unforgiving battery. When Colonel Albert Pope (1843-1909), one of the early builders, regaled a skeptical public with the prospects for the "storage battery. . . advancing more rapidly" than its competitors, he was thoroughly censured by Ingersoll. In fact, the acid reply branded Pope's assertion "an unqualified falsehood." ("A Discredited Expert," *The Horseless Age* 5 (October 18, 1899), 3.) Of course, Ingersoll's contention was ultimately corroborated, but Pope was surely being sincere at that initial stage.

Moreover, Ingersoll's skepticism appeared justified when news surfaced that two companies which had been founded years earlier on the premise of perfecting electric car batteries both went out of business in 1900. (Frank Oppel, ed., *Motoring in America: The Early Years* (Secaucus, New Jersey: Castle Books, 1989), 90.) The capitalization of the firms had combined for $50 million, while one remaining such research firm cut way back on funding.

[225] The commentator was one Ray Stannard Baker (1870-1946), a boisterous, outspoken, but far-sighted, pioneer author/journalist who early on appreciated the coming of the car. (Quoted in Phillip Van Doren Stern, *A Pictorial History of the Automobile* (New York: Viking Press, 1953), 22.) Years before this, he had written of the inevitability of the car, and hinted about the coming battle among the alternatives. (Ray Stannard Baker, "The Automobile in Common Use," *McClure's Magazine* 13 (July, 1896), 195-197.)

[226] When this writer checked through the dusty pages of contemporary magazines like *The Horseless Age*, the general consensus of the reporters was one in which all three main power types had a niche, a place in the automotive world. There are

numerous and sundry references to forthcoming advances in electric/steam car technology. So many different "advances" as to make for more interesting research, were one so inclined and had the time to do so.

Incidently, the writer had to rely upon these periodicals as primary sources to a greater degree than he had originally wanted to. In addition to a dearth of original pioneer accounts, "Business records have not survived for the lion's share of the approximately 1,500 firms. . . in the American automobile industry." Furthermore, records are, "virtually non-existent for the pioneer automobile dealers, mechanics, club officials, and owners." James J. Flink, *America Adopts*, 7. Under these troubling circumstances, legend and unsubstantiated fallacy often masks the facts. All of this truly handicaps serious attempts at historical research.

[227] Kimes, *Standard Catalog*, Third Edition, 44-45.

[228] Robert Q. Riley, *Alternative Cars in the 21st Century: A New Transportation Paradigm* (Warrendale, Pa.: Society of Automotive Engineers, 1994), 181, 184-185. Certainly a number of compressed air vehicles were in operation at this writing, but, according to Riley (1940-), the cost to retrofit a traditional service station to handle compressed air, "is expected to be in the range of $320,000." (ibid.) We looked at this situation merely to demonstrate the continued impracticality of this type for normal mass production use at this writing. As for the fuel, the problem was not the gas itself, but the cost of storing it when not in use. One big problem, according to Riley, was purchasing carbon— necessary for low pressure storage. In 1994, the cost of carbon was $20 per pound. By 2000, it was over $25 a pound.

[229] Floyd Clymer, *Treasury of Early American Automobiles* (New York: Bonanza Books, 1950), 7. The only thing left of this "car" is a drawing; this an undercarriage view.

[230] Bailey, et al, *The American Car*, 41.

[231] ibid., 40-41.

[232] Stern, 140. By the by, a connection with modern similar devices could not be traced. Certainly, though, efforts like this would never address the broader concerns of public transportation.

[233] ibid, 21.

[234] Apparently, this design utilized legs because an early notion believed that, when the wheels were given power, the only result, "would only cause the wheels to dig themselves in the ground." (Ken Purdy, *The Kings of the Road* (Boston: Little, Brown & Company, 1952), 195.) It did not take long to dispel this faulty perception.

[235] This gentleman and his brother, J. Frank Duryea (1869-1967), designed one of America's earliest cars.

[236] Nick Georgano, *The American Automobile: A Centenary 1893-1993* (New York: Smithmark Publishers, 1992), 12-13.

[237] Bailey, et al., *The American Car,* 52.

[238] For instance, what if the alternative cars had been given an adequate chance to survive, a "fighting chance?" Would gasoline cars have been driven into extinction thereafter? Would all three types have survived? Even, were the alternative cars given a chance after all?

[239] Stein, *The American,* 32. Anthony Bird's (1917-1974) similar premise seems sound. He contends the fact the steamers, "were barred from many [sporting] events. . . gave rise to the myths that. . . the steam car was 'killed by the opposition of the petrol companies.'" (Anthony Bird, *Antique Automobiles* (New York: Dutton Co., Inc., 1967), 130.)) We should note here, steamers were banned from these events because of the great lengths necessary to accommodate the type, *not* because they were inherently bad or dangerous.

[240] This, of course, is one of the chief arguments we are making here. To further discredit the 'conspiracy theory:' Ken Purdy, *The Wonderful World of the Automobile* (New York: Thomas Y. Crowell Co., 1960), 105, ("In those early days, before vast investments were made in refining plants, they [i.e., the petroleum interests] were as happy to sell kerosene as gasoline.") In fact, at the beginning of the auto age, "Kerosene was the main business of the oil people." (Musselman, 95-96).
Stern (1900-1984) gets more specific about that "other" petroleum product. "It [gasoline] was highly explosive when vaporized, difficult to store, and had no practical use." (Stern, *Pictorial History*, 14). That being the case, it is enlightening to report gasoline's low regard in the earliest pioneer days. "Gasoline, little regarded, was largely wasted. [But] it was also cheap and plentiful." (Nicholson, *Passenger Cars 1863-1904*, 3).

That latter feature did have its adherents. Early pioneer Sylvester Roper (1823-1896), "fired his boilers with liquid fuel—kerosene or that dangerous new liquid, gasoline." (Stein, *The American*, 30. That was in the 1860's.) Incredibly, once upon a time, Federal inspectors were even put in, "oil refineries to make sure that no gasoline was put in the kerosene." (John Bell Rae, *The American Automobile: A Brief History* (Chicago: University of Chicago Press, 1965), 14-15.) Conclusion: Little fear of a gasoline 'conspiracy.'

[241] Despite the above, though, there have always been people who believed in a big conspiracy. Individuals who see the plottings of evil men in a faceless, uncaring government and an early version of 'Big Brother,' trying to guide the lives of the people. Among them, once the competition had been stifled, the story goes, the conspirators then took steps to make sure gas cars were the only ones left. Nothing has, to date, been found anywhere to prove this.

[242] Musselman, 35. Freeman elaborated: "Each worked independently, without knowledge of the other's problems or help from the other's experiences." (16-17). The American tinkerers were particularly guilty in this respect, according to this theory. Actually, the number of motoring magazines from this "Golden

Age" of the written word alone would have guaranteed few, if any, advances in automobilia might pass unnoticed by the trade at large. Moreover, as James Flink offers, "One of the pervasive myths of automotive historiography. . . [is] automotive pioneers worked largely in ignorance of one another's accomplishments." (James J. Flink, *The Automobile Age* (Cambridge, Massachusetts: MIT Press, 1988), 13).

[243] Many references support this tenet, including Scott, "Does Mourning Become?," ("This brainchild of. . . Kettering['s] effectively sealed the doom of the electric."), 204; Frank Donovan, *Wheels for a Nation* (New York: Thomas Y. Crowell Co., 1965) ("the greatest single invention since the birth of the automobile [was] the self-starter"), 134; Stein, *The American*, Certainly, "The gas car's decisive weapon was the electric self-starter of 1912. . .", 46; Wager, after the introduction of the self-starter, "steam [was] just too much trouble to bother with," 64. *Encyclopedia Brittanica*, s. v. "Automobile," George C. Cromer, Orville C. Cromer, (Chicago: University of Chicago, 15th Edition, 1984), "[T]he electric self-starter. . . did as much as anything to doom the electric car by eliminating the dreaded hand crank and making the internal combustion engine car amendable to operation by women." (518.) Make no mistake; this is a very entrenched idea. But one which does not bear up under close scrutiny. Nevertheless, Rae, *American Automobile*, 48; Robert F. Karolevitz, *This Was Pioneer Motoring* (Seattle, Washington: Superior Publishing, 1968), 158-159; Richard Schallenburg, *Bottled Energy: Electrical Engineering and the Evolution of Chemical Energy Storage* (Philadelphia: American Philosophical Society, 1982) all have a similar viewpoint to the above cited sources. Schallenburg labels the introduction of the self-starter, "the nail in the coffin" for the electric car, and presumably for the steamer as well. (274). This is really difficult to contend since electric car market penetration, except at the start, was *always* so slight.
On the other hand, Kirsch (1964-   ) states, "It [i.e., the self-starter] worked and allowed the electrified [note the wording] gasoline car to make significant inroads into the already narrowing market niche for pure electric vehicles." (216). (David A. Kirsch, *The Electric Vehicle and the Burden of History* (New Brunswick, New

Jersey: Rutgers University Press, 2000).) Significantly, then, even dedicated electric car proponents like Kirsch will admit to this, "already narrowing market niche" of the days **before** the self-starter. All of which petitions the question, 'Why was the popularity of the alternative cars already slipping *prior* to the self-starter if they were so much better than their internal combustion rivals?'

[244] However, the faulty perception was not universal among the consulted sources. For instance, G. N. Georgano, *Vintage Cars: 1886-1930* (Twickenham, England: Tiger Books, 1977) ("The introduction of the self-starter had something to do with this [i.e., the decline of the alternatives], but people's expectations were rising as well."), 38.
We do not mean to imply for a moment the successful invention of a practical self-starter had **no** advantages. Part of the significance of this invention is in this statement, "The claims of contemporary journalists that the self-starter did as much for the emancipation of women as universal suffrage may have been sexist and exaggerated, but. . . [it] certainly helped put many more women comfortably behind the wheel." Moreover, in related fashion, "the self-starter opened up an enormous market among. . . [those] who saw the automobile primarily as a means for traveling with minimum fuss and maximum convenience." (Stuart W. Leslie, "Charles Franklin Kettering," 247. Quoted in *Encyclopedia of American Business History and Biography, 1920-1980* (New York: Facts on File, 1981).
There was another change. Before the self-starter appeared, "dashing teams of horses continued to draw most fire-engines." The explanation was a simple one. Gentlemen in this line of work could *not* take the required time for the steamer's start-up. Nor could the electric fulfill their special needs. After all, in this profession, time was of the essence. However, once the Self-Starter proved the worth of the 'gassers,' the 'dashing teams' of horses were literally put out to pasture. (Elizabeth Janeway, *The Early Days of Automobiles in America* (New York: Random House, 1956), 67. Finally, and most importantly, the appearance of the device made any comeback by the alternative cars highly unlikely. How fittingly ironic that the electric car was done in by an electrical component in the end.

However, none of the foregoing suggests the alternatives declined in the first place *because* of the Self-Starter. This point *must* be kept clear.

[245] One might at first glance assume production figures for the alternative *vs.* gasoline cars should be readily available. Such figures would be useful in trying to figure out how the war to win the hearts of Americas consumers played itself out in those early days. Unfortunately those records, assuming they were *ever* available, did not exist in written form at this writing. Inquiries by this writer of various printed early media, including Ward's Automotive, failed to discover them. Additional analyses, carried out at the NAHC, failed of results. The former American Automobile Manufacturer's Association, renamed the Alliance of Automobile Manufacturers in 1999, had sent its old files to the Detroit Public Library. There they were sifted through, without result. The writer then launched his own independant research to reconstruct those figures to the most reliable degree possible—See Table A.

[246] Douglas A. Wick, *Automobile History Day-By-Day* (Bismarck, North Dakota: Hedemarken Collectibles, 1997), 203.

[247] ibid., 378.

[248] Kimes, *Standard Catalog,* Third Edition, 1378. Following a thorough investigation of any and all concerns that could have made steam cars during 1912 betrays steam car production from all sources as approximately 630 units.

[249] See Table A. The table reveals some interesting contrasts. As late as 1903, electrics made up more than a quarter of all new car sales in America, while the steamers had 13% of the market themselves. Their combined 39% share marked the last time alternative output would rank so high. Then, the bottom fell out. By 1906, electrics had declined to just 5% of the market, while the steamers accounted for a mere 4%. Taking up the slack was the burgeoning demand for the internal combustion cars. Significantly

for us, none of the latter yet had practical self-starters. Were years from them, in fact.

[250] Scott, "Does Mourning Become?", 204. Ironically, 1912 also turned out to be the year total electric car registrations in the U. S. reached their all-time peak. 33,482 electrics were listed in the hands of various owners. (Crower and Crower, 518.)

[251] ibid., 204.

[252] Clearly, this one factor, albeit an important one, of ease of starting, was not enough of an advantage to offset those more serious shortcomings. Consumers obviously realized this. Lack of range, for instance, would continue to be a serious drawback. One not easily tolerated.

[253] "Long Distance High Speed Electric Passenger Vehicle is Here," *Central Station* 11 (March 1912), 268. The title of the article betrays the misplaced optimism. It might behoove us to look over a contemporary viewpoint on how the alternative cars were viewed. In 1901, in *Plain Facts About the Automobile,* Albert Clough stated sharply, "Steam. . . has been developed further by American builders than either of the other motive powers, unless possibly the electric." (Quoted in Fred J. Wagner, *Saga of the Roaring Road: A Story of Early Auto Racing in America* (Los Angeles: Clymer Publications, 1949), 34.) Scarcely a year after, in an about face worthy of a discredited politician, Clough wrote the recalcitrant driver of the period, "can give up his 'pipe-dream' of electrical traction, take his oil [ergo, gasoline] engine and put in on the carriage, and go where he likes and as far as he likes, and not be a slave to a wire." (Albert Clough, "The Real Object of Their Hopes," *The Horseless Age* 10 (August 20, 1902), 186.)

[254] Maxim, 46.

[255] ibid., 62. Of course, strictly speaking, this would apply to steamers as well as gas cars, except the former required lots of water and advanced engineering knowledge by the operator as well.

[256] George Pope was Albert's brother. He shared his brother's affinity for electrics.

[257] John Bell Rae, "Hiram Percy Maxim," *Encyclopedia, 1896-1920*, 327. In fact, one of the most oft quoted phrases attributed to Albert Pope was his, "Who would willingly sit over an explosion?" (Stephen B. Goddard, *Colonel Albert Pope and His American Dream Machines: The Life and Times of a Bicycle Tycoon Turned Automotive Pioneer* (Jefferson, North Carolina: McFarland & Company, 2000), 13.) Pope was referring here to the basic principles of the internal combustion engine. A common outlook, but one that was fast becoming *passe*.

[258] Pragmatically speaking, however, *most* automotive businessmen were neutral on this whole issue. Their primary commercial interest was in producing a marketable vehicle. Not specifically what powered such a car.

[259] Moreover, the dangers of 'backfire' and broken/sprained wrists from the hand crank increased proportionately with greater sized engines.

[260] Ken Purdy, *Motorcars of the Golden Past* (New York: Galahad Books, 1966), 30. The car's engine had been fiddled with to prevent it starting during the test.

[261] James Melton, with Ken Purdy, *Bright Wheels Rolling* (Philadelphia: MacRae Smith Company, 1954), 70.) Melton once said jokingly about steam cars, "I love 'em." Even this author, though, failed to buy into any kind of 'conspiracy theory.' In fact, Melton wrote decisively about the steam car people. "They were way 'head [sic], and they just assumed they always would be." (ibid., 79).

[262] ibid., 76-77.

[263] Moreover, the gas car's lag was a short-lived one indeed. For the 1915 model year, Cadillac introduced a 314 cubic inch L-Head V8. This new powerplant boasted many advanced features,

developing 70 bhp @ 2400 rpm's. It was likely intended to be the last word in engine development, at least up to that point. In contrast, Pierce Arrow, which still had its six-cylinder program, could coax its enormous 825 six up to just 101 bhp @ 1600 rpm's. The Pierce engine was 2.63 times larger, but made just 1.44 times more power. Clearly, then, the big, slow turning six cylinder engine was no mean match for the much more efficient, much smaller V8. The latter clearly had the better future. No wonder, "Among automobile engineers themselves, it was realized that Cadillac had produced a giant killer."
Moreover, the comments of W. R. Strickland, chief engineer at Peerless, about the introduction of the Cadillac engine are noteworthy. "The many advantages of the eight. . . led to [a] complete change of thought. Eights (and twelves) began to multiply and displace the older large bore engines." (Maurice D. Henry, *Cadillac: Standard of the World* (Princeton, New Jersey: Princeton Publishing Co., Third Edition, 1979), 106.)
This inevitable downsizing of internal combustion engines courtesy of new technology occurred with great impetus, once it got underway. "The fact of diminution of displacement coincid[ed] with increased power-speed potential." (Peter Helck, *Great Auto Races* (New York: Harry N. Abrams, 1975), 33.) Getting more performance from smaller engines!

[264] As a saving grace of sorts, electric motors as driving units were the least complicated of the three main types to make, thus necessitating less start-up capital for new firms entering the field. As a result, there were a sizeable number of electric car marques.

[265] *Floyd Clymer's Historical Steam Car Scrapbook*, 70.

[266] "Electric Automobiles," *Electrical World* 32 (November 5, 1898), 465-466.

[267] Admittedly, the earliest electrics employed a primitive *form* of a pneumatic tire, but, "the excessive weights of vehicles and loads caused frequent blowouts." (Scott, "Does Mourning?", 203). As a result, solid tires were then employed out of sheer

necessity. When the latter demonstrated they were prone to provide harsher riding qualities and lacked durability, the tire makers had to hastily come up with an improved, but again pneumatic, replacement. We have already observed Alvaro Krotz's experiments in this direction.

[268] ibid., 203.

[269] Charles Duryea, "The American Motor Car Industry," *Motor* 11 (March 1909), 35-123, inclusive. By the by, the Locomobile had another difficulty: the little steamer, *sans* condenser, was good, "for only about 20 miles on a tank of water." Http://www.hfmgv.org/histories/pic/97.sep.html. October 13, 2000. "Locomobile Steam Runabout."

[270] This was usually the type attracted to the fledgling car industry; such was the nature of the beast.

[271] Salom, for one, was a real proponent of the early electrics. He gave a scathing denouncing of the gasoline engine, which is found in *The Journal of the Franklin Institute*. "All the gasoline motors we have seen belch forth from their exhaust pipe a continuous stream of partially unconsumed hydrocarbons in the form of a thick smoke with a highly noxious odor. Imagine thousands of such vehicles, each offering up its column of smell." ("Stated Meeting, Dec. 18, 1895. Mr. Jos. M. Wilson, President, in the Chair: Automobile Vehicles by Pedro J. Salom," *The Journal of the Franklin Institute* 141 (April 1896), 278.)

[272] Ralph Stein, *The Treasury of the Automobile* (New York: Golden Press, 1961), 102.

[273] Minutes, Association of Edison Illuminating Companies, 1899, 197-206.

[274] "Electric Vehicles and Their Limitations," *The Horseless Age* 4 (September 27, 1899), 5. This, in retrospect, seems a

remarkably frank admission. In fact, it could have been uttered at the end of the twentieth century with equal veracity.

[275] These were necessary, even though many electric car makers provided customers with mobile do-it-yourself charging units. There was a wide variety of these "portable"devices. But they all had one common malady: they were all slow and cumbersome at best.

[276] Karolevitz, 117.

[277] Flink, 241. This would really make for an expensive proposition. Besides the other inconveniences, the electrics thus made even less sense from an economic standpoint. An extreme example was an electric car sent on a jaunt from Boston to New York City in October, 1903. Getting a recharge was not a problem, but the cost at various stations along the way, "was $15, or between four and five times the cost of gasoline for propelling a four passenger gasoline car the same distance." ("The Long Distance Record Run of an American Electric Automobile," *Scientific American* 89 (October 31, 1903), 310.)

[278] Thomas LaMarre, "Detroit's Electric: Society's Town Car," *Automobile Quarterly* 27, no. 1 (Fall 1989), 164-166.

[279] Wager, First Edition, 212. Such charging stations as did exist varied widely in their capabilities, and were generally found only in large eastern cities at that. However, had the electric car been more viable, charging stations in other parts of the country would have been forthcoming. But there was no incentive to build them. A Catch 22 situation at best.

[280] "Work of the Electric Hansom in New York," *Horseless Age* 2 (July, 1897), 12.

[281] Kirsch, 48.

[282] The Chicago branch packed it in when the "cabbies" called a strike over profit-sharing. It is not clear what caused

the drastic move, although overexpansion of the Pope-Whitney plan was probably to blame. One journal called this, "a discouragement rather than a surprise to those . . . watching the situation." ("Automobiles in Chicago," *Electrical World* 37 (March 16, 1901), 429.) Other branches endured similar fates, except the New York division.

[283] Stein, *Treasury,* 101-102. Hiram Percy Maxim, "Radius of Action of Electric Motor Carriages," *The Horseless Age* 2 (July 1897), 2-4. Maxim contended quick charges, over time, could greatly shorten useful battery life.

[284] Early batteries were much larger, bulkier, and heavier. Some weighed around 750 pounds, per each battery, although this was an average. A few weighed a great deal more. The appellation 'battery' could be a misleading one then.

[285] Henry Jackson Howard, "American Progress in Automobilism," *Metropolitan Magazine* 11 (January 1900), 5.

[286] Stein, *The American,* 55.

[287] Karolevitz, 9, 167; John A. Conde, *Cars With Personalities* (Keego Harbor, Mich: Arnold Porter Publishing, 1982), 8. Note, however, the builders of the Detroit Electric proudly boasted, "Thomas A. Edison chooses the Detroit Electric *exclusively* as the one car properly made to use efficiently the tremendous capacity of the Edison battery." (Robert Karolevitz, *Old-Time Autos in the Ads* (Yankton, SD: The Homestead Publishers, 1973), 65.) Of course, like so much in advertisement, this was patently untrue.

[288] Indeed, many early twentieth century women used the car as a vehicle for social advancement, and acceptance. Every bit as much as for transportation. In fact, "With her electric brougham, the American matron could go shopping and pay social calls[,] independent of husband or chauffeur." (Georgano, *American Auto, 1893-1993,* 71.) Interestingly, some women of

quality, "chose to operate the car from the rear seat so as not to give the impression they were doing such a lowly thing as driving." (Nelson Bolan, *Running Board Cars* (Lawrenceville, Virginia: Brunswick Publishing Company, 1987), 21.) The sight of a car coming down the street without a driver must have been intimidating to the uninitiated! Only electric cars could have been so treated. With the double-clutching necessary for gas cars in those days, there was no way to "drive" a gas car from the back seat. Or a steamer for that matter. (We will shortly deal with early women motorists and the alternatives.)

[289] We can not help but wonder what the result would have been had Whitney & Co. been able to sell the American public on the abstract idea of the selling of a service—i.e, taxi service on a mass scale rather than private cars—instead of what did happen. Of course, this would likely have been an exclusively urban phenomenon under any circumstances.

[290] Kirsch, 31.

[291] John Bentley, *The Oldtime Automobile* (Greenwich, Connecticut: Fawcett Publications, 1951), 108-109.

[292] James J. Bradley and Richard M. Langworth, "Calendar Year Production: 1896 to Date," *The American Car*, 138.

[293] W. H. Palmer, "The Storage Battery in the Commercial Operation of Electric Automobiles," *Electrical World* 39 (April 12, 1902), 646. In fact, much of the reason the conglomerate succeeded *then* was due to the fact it embarked on car production in-house. "Columbia and Electric [Vehicle Company] built some 2,000 electric cabs in the first large-scale production operation in. . . the industry." (John Bell Rae, "Electric Vehicle Company," *Encyclopedia, 1896-1920,* 174-175.) By the way, on June 20th, 1900, the EVC had acquired complete control of the Columbia & Electric Vehicle Company. This maneuver put them squarely into

the manufacturing business, not just functioning as a holding company. (*The Horseless Age* 7 (December 12, 1900), 14).

[294] Since historical events do not take place in a vacuum, we must remember the Sherman Ant-Trust Act, passed in 1890 to thwart a monopoly on transportation by the railroads, had become almost a battle cry by then against any entity that vaguely resembled a monopoly. The EVC sure fit that bill.

[295] Readers desiring further details are referred to John Bell Rae, "The Electric Vehicle Company: A Monopoly that Missed," *Business History Review* 29 (1955), 298-311. Rae (1911-1988) summed up by calling the EVC, "a parasitical growth on the automobile industry. . . [whose] demise was regretted only by those unfortunate enough to hold its securities." (p.311)

[296] Kirsch, 31.

[297] *Automobile* 17 (December 12, 1907), 881.

[298] George Baldwin Selden (1846-1922), a patent lawyer from Rochester, New York, filed for a patent on gasoline automobiles in 1879. Through clever legal wrangles, Selden was able to forestall the granting of his patent until 1895, just when the gasoline car was appearing in sufficient numbers to warrant his interest. He then broadly tried to claim his patent covered all internal combustion engines, so when the automobile took off Selden, in company with his new partners in the EVC, announced to the industry the patent was binding, and demanded royalty payments from all gas car companies. Now this was something of a gamble. In 1895, or even 1905, no one could have known how big the gas car monopoly would be.
None of this is meant to suggest Selden himself was less than astute. Thus, after looking over the 'heavy road carriages' of his day, "as early as March, 1873, he definitely rejected the idea that a steam carriage could satisfy those [transportation] requirements." (William Greenleaf, *Monopoly on Wheels: Henry Ford and the Selden Automobile Patent* (Detroit: Wayne State University

Press, 1961), 11-12.) Moreover, the ripples of the permit were felt both far and wide. To comply with the charter, several manufacturers did join together to form the American League of Automobile Manufacturers (ALAM), but others refused to go along. Henry Ford finally defeated the patent in the courts in 1911, ending the monopoly. (William McGaughey, *American Automobile Album* (New York: E. P. Dutton, 1954), 65.) Ironically, this very famous patent (#549,160, for the record) has been, to date, the only attempt to corner the market on *any* of the motive power sources the car had available. Selden's apparent vision was wrong on one count, though. His idea revolved around the two-cycle engine design. This arrangement, outside of a few proponents like the Elmore Company of Clyde, Ohio, never proved popular with American manufacturers.

[299] There were other contributing factors. In January, 1907, two hundred electric cabs tied up at a large urban charging station were irreparably damaged by a debilitating fire. ("Hundreds of Electric Cabs Burned," *Power Wagon* 3 (April, 1907), 4.) To stay in business, the company was then compelled to turn to gasoline cabs. By 1909, domestic gas cabs made up fully one-half of the New York City fleet. (Joseph Anglada, "The Taxicab Business in New York," *The Horseless Age* 24 (July 21, 1909), 63-65.)

[300] Karolevitz, *Pioneer Motoring,* 159. In a letter to the writer, dated October 19, 2001, Mr. Karolevitz expressed his regret he could not provide further information. He could not elaborate on how the story had arisen in the first place.

[301] Henry, 84.

[302] George S. May, "Cartercar," *Encyclopedia, 1896-1920,* 75. May mentions the story, but quite justifiably rejects it as a myth.

[303] Northcoate Hamilton, "Self Starters," *Encyclopedia, 1896-1920,* 415. This is some careless editing on someone's part.

[304] Nick Baldwin, G. N. Georgano, Michael Sedgwick, Brian Laban, *The World Guide to Automobile Manufacturers* (New York: Facts On File, 1987), 86.

[305] A wide variety of devices had been attempted, with varying results. Some, for instance, were very complicated. The McFarlan, a prestigious limited production car, utilized a compressed air starter of the company's own design. This was standard on all 1912 models. The starter was operated by, "a four cylinder Kellogg pump which stored air under 200 lbs. Pressure." (Keith Marvin, with Alvin J. Arnhem, and Henry H. Blommel, *What Was the McFarlan?* (New York: Privately Published, 1967), 8.) After Kettering's model hit the market, though, these other designs were quickly abandoned as impractical. (Gregg D. Merksamer, *A History of the New York International Auto Show: 1900-2000* (Atlanta, Ga.: Lionheart Books, Ltd., 2000), 38.)

[306] Both gasoline and kerosene were then usually sold out of hardware stores. This was one of the ironies of the whole situation. Although gasoline was plentiful, getting a supply of it could be difficult. Few wanted or needed it. (See Footnote # 239) It might be worth noting the petroleum industry was much smaller and less affluent at the beginning of the twentieth century than it was at the end. Kerosene was used primarily as a lamp oil and gasoline's chief, almost only, use was as a laundry additive. It was true enough steamers needed some source to heat the steam; but kerosene seems to have been the fuel of popular choice. It was much safer and easier to handle than the more volatile gasoline. How dangerous was the latter? "The fuel of 1900 was about equal in volatility to a modern [*circa* 1965] high-test gasoline." (Donovan, 133.)
Interesting, once the gas car took off, the demand for its fuel skyrocketed. In 1911, demand for gasoline overhauled kerosene for the first time. This whole situation and the evolving sense of gasoline's worth provides a microcosm of society's changing values. (Karolevitz, *Pioneer Motoring*, 106-107.) In a remarkably short span of time to boot.

[307] Maxim, *Horseless Carriage*, 6. In fact, Maxim left no doubt of his allegiance. With typical understatement, he wrote: "In the year 1900 the much-despised gasoline carriage... began to give indications of establishing an important industry in spite of its 'greasy machinery.'" (ibid., 169).

[308] Ironically, this rugged pioneer never seems to have taken to the early electric buggy in the same manner that most early women motorists did.

[309] Janeway, 144-145. Apparently a Peerless car was stopped in her path, and when she swerved to avoid the vehicle, she and a couple of traveling lady companions went right over the side. After receiving help extricating her ride, the jaunt continued.

[310] John A. Heilig, "The Glidden Tours," *Automobile Quarterly* 30, no. 3 (Spring, 1992), 19.

[311] Janeway, 145.

[312] Heilig, "Glidden Tours," 19.

[313] Her first car, bought in July, 1902, was a little steamer she drove all over New York City. Much to her chagrin, the boiler burned when she left the burner control on one day while parked. Later, she bought the White. (Robert Sloss, "What a Woman Can Do With an Auto," *Outing Magazine* 56 (April, 1910), 236ff. By the way, what *Mr.* Cuneo thought of all of this is not recorded.

[314] ibid.

[315] Sloss, 236ff.

[316] Karolevitz, *Pioneer Motoring*, 158.

[317] Ruddock, 35. Freeman dubbed them, "floating greenhouses." (Freeman, 123.)

318 Keith Marvin, "Off the Beaten Track. . . Being a Study of the Beautiful and the Bizarre: Latter Day Electric in all its Glory," *The Upper Hudson Valley Automobilist* 34, no. 1 (January 1984), 13.

319 Stern, 242.

320 Georgano, *American Auto, 1893-1993*, 70-71.

321 William T. Cameron *The Cameron Story* (Tuczon, Arizona: International Society for Vehicle Preservation, 1990), 15-17.

322 We should note, though, not all steamers utilized such features. The Grout Steamer, for instance, used, "a specially designed pilot [that] needed no spirits or irons to ignite." (Stanley K. Yost, *They Don't Build Cars Like They Used To* (Mendota, Illinois: Wayside Press, 1963), 103.)

323 The writer here summarized the starting procedures as found in Stein, *Treasury,* 103-105; Georgano, *Veteran Cars,* 30, etc. You could drive once this was done, after you released the parking brake. Until the water ran low. For most steamers, this was often.

324 However, the economics were, at least, similar to the internal combustion cars. "The monthly cost of maintenance should not exceed $25," for both steamers and gasoline cars.("Automobile News," *Scientific American* 88 (May 9, 1903), 354.)

325 By the way, electric car manufacturers could here show a distinct advantage. Auto maker Walter Baker (1867-1955) could not resist pointing out the electric car, "leaves nothing to freeze, burn, or explode." (Karolevitz, *Old-Time Autos,* 46.) Baker built his electrics in Cleveland.

326 Jamison, 67.

[327] ibid., 67-68.

[328] Flink, 236.

[329] As one writer stated about these barren regions, "Any use of steam automobiles. . . [there] would have necessitated service stations supplied with water brought in from distant points." (Rae, *American Auto,* 14.) Incidently, the idea of an external condenser that could strain water to remove impurities was a good one, but not universally accepted. (E. J. Stoddard, "Water for Automobile Boilers," *The Horseless Age* 7 (December 26, 1900), 18-19.) It further complicated what was already a complex system.

[330] Maxim, *Horseless Age,* 37.

[331] Kirsch, 266; John F. Katz, "F. E. & F. O. Stanley: The Challenge from Steam," *Automobile Quarterly* 25, no. 3 (Spring 1987), 41.

[332] To be fair, there *were* a few isolated incidents. In May 1902, a Mr. F. W. Chase of Worcester, Massachusetts, and four others were injured when the gasoline tank of his Chase Steam car blew up while, "the inventor was preparing for a trial run." (Kimes, *Standard Catalog,* Third Edition, 279.) During the 1901 Madison Square Garden Auto Show, a Detroit investor named Henry Joy and a friend, Truman Newberry, were, "inspecting a steam-driven Locomobile when a glass tube indicating the boiler water level exploded within three feet of Newberry's face." Newberry was fortunate to escape serious injury. (Merksamer, 13.) Joy forthwith turned his interest to gasoline cars, and subsequently became president of Packard Motor Car Company.
Incidents like these, although isolated and rare, did lend credence to public fears about steamers. Especially when the motoring press of the day made sure such stories were circulated far and wide. We have already observed the electric car's negative image

in the pages of *The Horseless Age*. It was not long before there was a general consensus of opinion to damn its steamer cousins in contemporary motoring magazines as well. In fact, if there was any feasible link to a so-called "gasoline" conspiracy at all, it was in the motoring press of the day.

[333] David Cohn, *Combustion on Wheels* (Boston: Houghton Mifflin Co., 1944), 131.

[334] Reginald M. Cleveland, S.T. Williamson, *The Road is Yours!—The Story of the Automobile & the Men Behind It* (New York: Greystone Press, 1951), 186.

[335] In fact, the English had really grown weary of steamers early on. Andrew Jamison (1949-) makes the case development of the type ground to a virtual halt in nineteenth-century England following an unfortunate accident. "In 1840 a boiler exploded. . . killing five passengers [on a private coach] and injuring twenty others." (Jamison, 39.) That resulting publicity, along with the infamous "Red Flag" law of 1865—passed as a means of placating 'Special Interest' groups like carriage makers/drivers—helped cripple car development in that country for decades. And we can not ignore improving boiler designs that arrived after the 1840 incident.

[336] Melton, 72.

[337] If that were not bad enough, the public were also, "thoroughly scared by the roar and hiss of surplus steam blown off in traffic hold-ups." (Peter Roberts, *A Picture History of the Automobile* (London: Triune Books, 1973), 71). Finally, there was, along with everything else, "motorists' fear of. . . accidents from the open fire maintained under the car." (Thea Bergere, *Automobiles of Yesteryear* (New York: Dodd, Mead & Co., 1962), 20. Ken Purdy (1913-1972) contends the latter feature, "kept steam cars out of public garages and off ferryboats unless their drivers shut everything down." (Purdy, *Kings,* 205.) We should note, though, early steamers, in retrospect, were really

relatively safe. Despite everything passed around about them by their detractors.

The Stanley brothers made a legitimate effort to determine at just what pressure their boiler would explode, if any. A guinea pig, identical to the production models, was placed in a pit. From a safe distance, we are not told how far, they dramatically increased boiler pressure until the boiler was leaking. It was holding 1500 pounds of pressure then. "The boiler wouldn't [sic] explode because every boiler tube had sprung a leak and had become a safety valve." (Janeway, 55-56.) It seems the brothers, in typical Yankee ingenuity, had wrapped piano wire around the boiler to, "handle the higher pressure." (James Laux, *Encyclopedia*, *1896-1920*, 426.) Obviously, at least in the case of the Stanley Steamer, sufficient precautions had been taken to ensure a safe product. Moreover, this was also a feature of the production models emerging from final assembly.

Unlike many more "modern" businessmen, the Stanley brothers appeared to know the product they were offering the buying public quite well. As another example, as early as 1897, the brothers were trying to perfect an Automatic Control to provide sufficient steam. (Richard and Nancy Fraser, *A History of Maine-Built Automobiles*, *1834-1934* (Privately Published, 1991), 14-15)).

[338] Joseph J. Schroeder, Jr. *The Wonderful World of Automobiles 1895-1930* (Chicago: Follet Publishing Co., 1971), 18. The source was a steam car advertisement of all things.

[339] ibid., 18. The ad implied, however, that buyers of *other* steamers besides the Victor would probably still experience anxiety anyway.

[340] There were still bound to be unpleasant scenes in this new age. On September 9th, 1901, in New York City's Plaza Square, 49 gasoline and 26 steam cars had lined up for the New York-to-Buffalo Endurance Run. Cars were becoming safer, even then. But human error had its place even then. At Peekskill, the designated stop for the second night, a small boy peered over into a parked steamer, and fiddled with the control lever. Steamers, as

we have seen, were usually left running and controlled by a single lever. The occupants were at dinner, so "steam was still up." The car, "promptly moved forward and ran over a twelve year old girl." (McGaughey, 26).

341 *Floyd Clymer's Steam Car Scrapbook,* 63.

342 ibid., 55.

343 *Floyd Clymer's Historical Motor Scrapbook,* Vol. 2 (Los Angeles: Clymer Publications, 1944), 64.

344 *Floyd Clymer's Historical Motor Scrapbook,* Vol. 5 (Los Angeles: Clymer Publications, 1948), 113. This ad appeared in 1912; by this time the gas cars were already firmly entrenched.

345 Marvin, *Cars of 1923,* 41. On the other hand, the same firm hedged its bet by continuing its usual line as well, albeit in a more abbreviated form. These efforts were not entirely futile. In 1916, when the company built 3000 cars, "half of all electrics registered in the entire country were produced by the Anderson Electric Car Company [makers of Detroit Electric cars.]" (LaMarre, "Detroit Electric," 165.) Unfortunately, this particular firm was forced to raise its prices in later years because of discontinuing lower-priced, less-profit cars, like the roadsters. These latter were dropped because too few sold to justify the costs of building the special types. It was a vicious, ultimately debilitating, cycle.
Nor was this the only firm to model their cars after gas cars. The Babcock Electric had, "what looked like a radiator filling spout." (Bellamy, 211.)

346 *Floyd Clymer's Scrapbook,* vol. 2, 106.

347 Bolan, 15.

348 At least according to this theory. While we are at it, there is an excellent one-page summary of electric car ads found

in Ruddock's article. It is entitled, "Winning the Female Vote." (41).

  [349] Desperate times often dictate desperate measures. In 1914, the New York City-based Electric Vehicle Association of America ran a full-page ad in, of all things, *Theatre Magazine*. In oft-repeated words echoing the old tried and true virtues of electrics, the placard almost pleadingly begged its readers, "The simple, safe, silent Electric is run by electric current–constantly decreasing in cost. . . Before you buy any car—consider the Electric." (Ruddock, 40.) Alas, too few potential buyers were listening by then. In fact, that argument had become a bit trite. In addition, by that point in time, the electric's disadvantages were also well-known. Within a few years, the association itself was disbanded. As the fortunes of the electrics waned, so did the organizations that promoted the type.

  [350] It could not have been because the White Steamers were inherently bad. LaMarre mentions a 1910 model owned by one Harry J. Sewell of Oakland county, Michigan. In 1930, the car was still faithfully running; having required no more than the usual basic maintenance. In fact, Mr. Sewell's picture appeared in *The Detroit News* of October 12th, 1930, under the caption: "He operates [the car] occasionally to give his friends a thrill." {LaMarre, "White Steam," 14.)

  [351] Shacket, 13.

  [352] Ruddock, 36.

  [353] Stein, *The American,* 108.

  [354] Kimes, *Standard Catalog,* Third Edition, 605. Styling was the one major area where, "the Franklin car. . . was having difficulty in competing with water-cooled vehicles." (Sinclair Powell *The Franklin Automobile Company* (Warrendale, Pennsylvania: Society of Automotive Engineers, 1999), 179-180.)

[355] Franklin's new passenger car registrations in the 1925 calender year were 7157, somewhat better than 1924's total of 5882. (Jerry Heasley, *The Production Figure Book for U. S. Cars* (Osceola, Wisconsin: Motorbooks International, 1977), 163.)

[356] *Floyd Clymer's Historical Steam Car Scrapbook,* 70. To put this in perspective, in one source, a contemporary photo showed cars traveling at an obviously stately pace. This was in, "1904[,] when 12 mph was considered good speed." (Stern, 90.)

[357] Such was the nature of this whole menagerie the very name of the racer is in dispute. Timothy Nicholson (1922-1991) states, in 1903 there appeared, "the first of the streamlined cars known variously as *Wogglebug, Teakettle, Rocket[,]* and *Beetle.*" (T. R. Nicholson, *Racing Cars and Record Breakers: 1898-1921* (New York: MacMillan Co., 1971), 121.)) These were all Stanley racers; Nicholson identifies Marriott's ride as the *Rocket.*

[358] Simply put, it had four power strokes for each revolution of the crankshaft.

[359] Raymond Walker Stanley, "Evaporating the Stanley Steamer Myth," *Automobile Quarterly* 11, no. 2 (Winter 1968), 123.

[360] Nicholson, *Racing Cars,* 121. We need to qualify these hp ratings, the measured rates of which appear too low. John Katz explains the Stanley brothers *may* have calculated their horsepower ratings the same way a nineteenth-century engineer would: "From the weight of water a boiler could evaporate in an hour." Thomas Marshall, a long-time Stanley enthusiast and steam engine buff, "independently estimated that the average Stanley engine could actually produce three to five times its rated power." (Katz, 21.) Nor was this situation unique to the Stanley. The White Steamers were, "well made, reliable and capable of developing 150 bhp in the 18 hp model over short distances." (Baldwin, et al., *World Guide,* 521.)

[361] Floyd Clymer, *Treasury of Early American Automobiles, 1877-1925* (New York: McGraw-Hill Books, 1950), 22.

[362] His name would subsequently appear on a famous line of motor vehicles.

[363] There exists in the NAHC a short paper entitled *'V, as in Victor, V, as in V8,' Footnotes to a History of the Racing Darracqs,* by Griffith Borgeson. Borgeson (1918-1997) did a thorough job on this paper. Through the kindness of the staff, the writer was able to obtain a copy of this work, which had been penned for the original curator of the collection, James Jerome Bradley (1922-1980). The racer's engine was a 90 degree V8, "bolted to a common, compact crankcase and using just one four-cylinder crankshaft and camshaft." Alas, all of the potential power of this beast seemed wasted at first. Chevrolet got his chance when Darracq team leader Victor Hemery was disqualified from the competition. It seemed he got rather incensed when the speed equipment failed to function, and, "his outrage in this case was expressed in the strongest language of [the] ex-Marseillais sailor which he was." (ibid., 8).

[364] Http://www.exford.co.uk/Steam/1906.htm. September 19, 2000. "The 1906 Stanley Land Speed Record." Interview of Fred Marriott, December 1955, by Ted Koopman, Newton, Mass. Mr. Marriott was interviewed when he was over eighty. He noted whimsically about his opponents from the day of the meet: "The look on their faces was sad and they couldn't [sic] figure out how the Rocket could make such fast time." That was a common occurrence among early steam car opponents.

[365] John Bentley, *Great American Automobiles: A Dramatic Account of their Achievements in Competition* (New York: Bonanza Books, 1957), 327-334. "The *Beetle(?)*, only 5 pounds lighter than its 2,200-pound rivals, ran away with the race until 200 yards from the finish, when it ran out of steam." 329. Lack of range was a common malady. Said one source, about the White Steamer—the only steam car to run in the Vanderbilt

Cup Races—it was, "totally unfit for such." (Helck, 177.) The Vanderbilt was a lengthy, endurance race.

[366] In yet another rather ironic twist, Marriott's accomplishment was never officially recognized because, "the run was not carried out under A.I.A.C.R. [Association Internationale des Automobile Clubs Reconus] rules." (Nicholson, *Racing Cars,* 13, 121.) Nonetheless, the "record" held until 1910. That was in spite of numerous attempts to better the times. The famous Mercedes of Berner Eli "Barney" Oldfield (1877-1946) nicknamed "Lightning Blitz," broke the record that year at 131.72 mph. Incidently, that car was another perfect example of our 'Age of the Giants,' boasting a mammoth overhead valve four cylinder of impressive dimensions. A 7.28345 in. bore X 7.874 in. stroke for its massive pistons gave 1312 cubic inches. (Wagner, 21, 80, 86-87.) Did it take Goliath to slay David this time around?

[367] Musselman does not indicate whether he believed the rumors; he was just seven at the time. (Musselman, 17.)

[368] "The grip of the steam automobile on the American imagination has been strong ever since[, so much so]. . . in the 1960's there were still 7,000 steam cars in the United States, about 1,000 of them in running order." (Cromer and Cromer, 515.) Considering the fact total steam car production never remotely approached the internal combustion cars, and that the last steamers were offered in small numbers in the early 1930's, this was a remarkable survival rate.

[369] Http://www.parkway-jars.com/cars/s0002.htm. April 14, 2002. "1910 Stanley 20hp Model 70."

[370] Stein, *Treasury,* 110. This intense pressure posed no apparent hindrance.

[371] Transcript of Marriott interview, page 3. By the way, this event was a total "steam" affair. Internal combustion rivals, doubtless wise from past experience, were cautious this time

around. In fact, "the Speed Carnival," was, "boycotted by most of the gasoline fraternity." (Helck, *Great Auto*, 177.) The electric cars kept a wary distance, too.

[372] Recorded in Musselman, 98. This was due, in part, to Marriott's unprecedented, "flying start–nine miles." (Purdy, *Kings*, 201.)

[373] Stein, *The American*, 37.

[374] Stein, *Treasury*, 110. Bentley, *Oldtime Steam*, 36, "A twisted scrapheap was all that remained of 'Beetle'. . . [after it] flew 100 feet through the air at almost 190 mph." Incredibly, even the entry on the Stanley brothers in *The Encyclopedia of American Business History 1896-1920* cites the terminal speed of Marriott's racer as the same patently erroneous 197 mph on that fateful day. (p. 425.)

[375] Purdy's *Kings*, for example, states, "the needle was passing 197 and still climbing fast when the car hit a little bump in the beach." (201.)

[376] A simple physics calculation can demonstrate the improbability of this happening. A terminal speed of 197 mph means 3.28 miles are covered for every minute of real time. With the Stanley's 34 inch tires, and the 1:1.75 gearing, this means the tires would have been spinning at 1944 rpms when the racer crashed, while the engine was turning at 1111 rpms. These figures are all approximations.
In retrospect, Marriott's enthusiasm with his racer probably clouded his judgement in later years, as Ray Stanley has suggested (Stanley, 128.) The car being geared in the manner it was meant, of course, the engine could only turn so fast. Specifications laid out by Marriott put engine rpm's as, "350 revolutions to the mile," with the wheels, "making 600 revolutions to the mile." (*Floyd Clymer's Historical Steam Car Scrapbook*, 70.) Finally, the calculations of Francis Stanley himself can not be discounted. The 1906 racer had had a new boiler installed by him and the, "valve

timing revised for better breathing." Then, Stanley anticipated the capabilities of the car with his improvements. He estimated it, "could now [before the fact] turn a mile in 25 seconds, i.e., at 144 mph." (Katz, 24.)

[377] Jamison, 156-157. Looking over all available data on the 1907 incident, and taking all factors into account, strongly suggests Dr. Ayres' assessment was the correct one.

[378] Beverly Rae Kimes with Henry Austin Clark, Jr. *Standard Catalog of American Cars 1805-1942* (Iola, Wisconsin: Krause Publications, First Edition, 1985), 1298.

[379] Stanley, "Evaporating," 128. This is clearly another of those misconceptions/errors we have already mentioned. Or at least mysteries. As such, the tale shares a few "red flags" with the Belle Isle incident: 1.) The professors are not identified, so the veracity of the report can not be verified; 2.) The means by which the alleged speed was arrived at has been left out; no doubt, intentionally.
Paul Hayes, writing in 1957, mentioned that, "Two Massachusetts Institute of Technology professors at the timing stand reported his [Marriott's] speed was 190 mph at the half-way point." (Paul Hayes, "The World's First Rocket: The Stanley Steamer World Speed Record Racer," Reproduced from *Modern Man Quarterly* 1957, From Http://www.exford.co.uk/Steam/Articles/mmql.htrr. September 9, 2002.). Moreover, this report said that in 1940 professors from the Massachusetts Institute of Technology (MIT) had arrived at a terminal speed of 190 mph; again, as Ray Stanley points out, "By what means or formula they arrived at such a conclusion, if there ever was one, was not disclosed." (128.)

[380] Musselman, 98.

[381] Griffith Borgeson, *The Golden Age of the American Racing Car* (New York: Bonanza Books, 1966), 34.

[382] Bentley, *Oldtime*, 34.

[383] At the risk of looking at purely American achievements, the honor of first traveling more than a mile in a minute goes to an electric car.

[384] Stein, *Treasury,* 100. Jenatzy's ride was specially constructed as a record breaker; "Electric motors were geared directly to the rear axle." (Nicholson, *Racing Cars,* 110.) This car bore a remarkable resemblance to an egg on wheels.

[385] ibid., 100-101. Worse, Bird said these runs were even more debilitating to the batteries because they, "sulphated from too violent discharging." (Bird, 130.)

[386] Purdy, *Motorcars,* 94.

[387] Wager, 206-207. Curiously, just before Baker's "record-setter" on the same day, a boiler car driven by one S. T. Davis set a new steam car record in the mile. The run was a mile in one minute, 12 seconds. This was, "3 seconds better than the previous record." (Scott, "Does Mourning?," 199).

[388] ibid., 207.

[389] Nicholson, *Passenger Cars 1863-1904,* 144. Paradoxically, Baker's efforts probably did little more than demonstrate the electric car's continuing impracticality. It displayed, in no uncertain terms, the accommodations necessary for all battery cars. Not the least of these being a hefty price tag.

[390] Which we will look closer at in a few moments. The Car actually employed two 1 ½ hp Lundell motors, one mounted on each drive wheel.

[391] Flink, *America Adopts,* 42. By the way, the winner's average speed was 26.8 mph.

[392] Bentley, *Oldtime Steam,* 48-49.

393 High pressure cylinder of 3 in. Bore X 3.5 in. stroke, with low pressure cylinder of 5 in. X 3.5 in. stroke. Total swept volume was 187 cubic inches.

394 LaMarre, 24.

395 Wager, 58.

396 With a T-Head six of 5.38 in. bore X 5.5 in. stroke for 751 cubic inches.

397 Clymer, *Treasury*, 43.

398 Early Minnesota Auto Racing, Contact Stauffer. No date. Available: <http://www.home1.gte.net/stauffer/Minnrace.htm>. The sight had a couple of good photos of "Whistling Billy."

399 Marvin, et al., *What. . . McFarlan?*, 4-5.

400 Cohn, 60-61; Musselmann, 138.

401 Musselmann, 125.

402 Riker's electrics were heavy, as a rule. For instance, a 1900 Riker Electric Brougham, a four-passenger car, tipped the scales at around 4000 pounds. (Clarence P. Hornung, with James J. Bradley *100 Great Antique Automobiles* (New York: Dover Publications, 1991), xvii.)

403 David Burgess Wise, *Classic American Automobiles* (New York: Gallahad Books, 1980), 25.

404 Making it, in the process, the number one selling car in the country.

405 Henry Ford Museum& Greenfield Village, Contact Molly Zink, *Present Old 16*. No date. Available:<http://www. old16.ORG/HTdocs/old 16/INDEX.HTML> This race car was in the possession of the great automobile artist Peter Helck (1893-

1988) until his death. After that, it became the property of the museum.

[406] Wagner, 8.

[407] The Editors of Automobile Quarterly, *GM: The First 75 Years of Transportation Progress* (Princeton, New Jersey: Princeton Publications, 1983), 15.

[408] Russell H. Anderson, "First Automobile Race in America," *Journal of the Illinois State Historical Society,* 47, no. 4 (1954), 343-360. Readers interesting in learning more about this particular race are referred there.

[409] We might note the pronounced difference in weight of just two of the entries: The Morrison-Sturges entry weighed a hefty 3535 pounds, *sans* passengers, while the more svelte Duryea tipped the scales at a mere 729 pounds. (Richard P. Scharchburg, *Carriages Without Horses: J. Frank Duryea and the Birth of the American Automobile Industry* (Warrendale, Pennsylvania: Society of Automotive Engineers, 1993), 109.)

[410] Peter Helck, *The Checkered Flag* (New York: Castle Books, 1961), 40.

[411] There was another, not intangible, results from all of these competitive events. "The tour [the 1905 Glidden] has proved the automobile is now almost foolproof. . . [and] it has strengthened our belief in the permanence of the motorcar." (George O. Draper, "A View of the Tour From One Participating," *The Horseless Age* 16 (July 26, 1905), 153.)

[412] Roberts, 73.

[413] Stern, 126. "The electric car presented far fewer problems than those powered by steam or i.c. [internal combustion] engines. . . [except for] its limited range between battery charges." (Marco Matteucci, *History of the Motor Car*

(New York: Crown Publishers, 1970), 44.) This really came in handy in days before garages were commonplace.

[414] Cohn, 45.

[415] Conveniently, the early women's movement was more or less contemporary. So the two factors were intertwined, although outside the parameters of this paper.

[416] "The Electric Cab Experiments," *The Horseless Age* 4 (August 30, 1899), 7.

[417] Karolevitz, *Old-Time Autos,* 27. Henry Ford received some sage advice from one-time boss Thomas Edison about his experiments with internal combustion cars. Around 1900, Edison told Ford, "You have the right idea." In like vein, "this car has an advantage over the electric car because it supplies its own power." (Peter Collier and David Horowitz, *The Fords: An American Epic* (New York: Summit Publishing, 1987), 34.) Curiously, Ford worked for Edison's electric concern but seems never to have experimented seriously with electric cars.

[418] William A. Cannon, Fred K. Fox *Studebaker: The Complete Story* (Blue Ridge Summit, Pennsylvania: Tab Books, 1981), 26-27.

[419] "Car traffic rose from 8000 vehicles in 1900 to 2,300,000 in 1915." This amounted to a 288% increase. (Enzo Angelucci and Alberto Belluci, *The Automobile From Steam to Gasoline* (New York: McGraw Hill, 1974), 119.)

[420] Peter Roberts, *Collector's History of the Automobile* (New York: Bonanza Books, 1978), 60.

[421] Ralph Stein, *The World of the Automobile* (New York: Ridge Press, 1973), 85. Loud, often raucous noise, producing lots of blue or black smoke.

[422] James M. Flammang, et al., *100 Years of the American Auto: Millennium Edition* (Lincolnwood, Illinois: Publications International, 2000), 30; Beverley Rae Kimes, et al., *Packard: A History of the Motorcar and the Company* (Princeton, New Jersey: Princeton Publishing Company, 1978), 32-33.

[423] John Robert Day, *The Bosch Book of the Motor Car: Its Evolution and Engineering Development* (New York: St. Martin's Press, 1976), 104-105.

[424] Flink, *America Adopts,* 281; Maxim, 121.

[425] Sedgwick, *Early Cars,* 87.

[426] Cohn, 40.

[427] ibid., 40-41.

[428] Rae, *American Auto,* 15. In effect, then, the gasoline car enjoyed a formidable 166.66% (167%) greater rate of efficiency than its nearest competitor, according to the above source.

[429] Nicholson, *1863-1904,* 112-113.

[430] Jamison, 40.

[431] Roberts, *Picture History,* 69.

[432] Flink, *The Automobile,* 7.

[433] John Robert Day, *The Bosch Book of the Motor Car: Its Evolution and Engineering Development* (New York: St. Martin's Press, 1976), 85. This particular factor remained, even at this writing, a most serious drawback to the widespread use of steam power for automobiles. The general trend of the auto industry in recent decades had been to find more efficient means of propulsion. The steam engine obviously requires *more* to do the same tasks than its internal combustion rival, hardly an improvement.

[434] Abner Doble, "Steam Cars, Past and Present," *Automobile* 26 (November 2, 1916), 781. The implication being, that if the perfect steamer could be built, then Doble might be in the best position to build it.

[435] James M. Laux, "Steam Cars," *Encyclopedia, 1896-1920,* 427. Certainly, this kind of prestige had its advantages. No less a personage than the reclusive Howard Hughes (1906-1976), when asked about prestigious American cars, said, "I know them all, but I prefer the Doble." (The Editors of Automobile Quarterly, *World of Cars* (New York: Bonanza Books, 1981), 104.

[436] G. Marshall Naul, *The Specification Book for U. S. Cars 1930-1969* (Osceola, Wisconsin: Motorbooks International, 1980), 106-107. "Prices ranged between $8000 and $11000 with a ready buyer for each car the factory turned out." (James P. Moloney *Encyclopedia of American Cars 1930-1942* (Sarasota, Florida: Crestline Publishing Co., 1977), 376.)

[437] Kimes, First Edition, *Standard Catalog,* 206.

[438] Abner's goals were nothing less. In company with his three brothers, he set out to show up the "gasoline crowd." "The Dobles were not introducing just another car, they intended to reassert the superiority of steam itself." (*World of Cars,* 104.) They failed through no fault of their own. For even the ***best*** of the steamers still had crippling flaws.

[439] That pressure was, "1,250 pounds per square inch." This allowed the 75 hp 4 cylinder steam engine to, "propel the 3,900-pound coupe in which it sat to 90 miles an hour." (Leon Mandel, *American Cars: From Harrah's Automobile Collection* (New York: Stewart, Tabori & Chang, 1982), 96.)

[440] However, even Doble could not solve the freezing problem. He tried mixing alcohol with the water to, "lower the freezing point." (Jamison, 49.) The high cost of the alcohol made this remedy impractical for mass consumption, though.

[441] Thomas Derr, *The Modern Steam Car and its Historical Background* (Los Angeles: Clymer Publications, 1944), 130-132.

[442] Stanley K. Yost, *The Great Old Cars—Where Are They Now?* (Mendota, Illinois: Wayside Press, 1960), 99.

[443] Since steamers were already complicated, their prices were *always* high, even at the very beginning.

[444] 1920 Stanleys bore a close resemblance to gas car competition; even boasting a 'radiator filler cap.' After the death of his twin in a car wreck, Freeland ceased having anything to do with his own company. By 1923, the firm was in the hands of a receiver, asking for protection from creditors. "Its plants and assets [were] sold early the following year for $572,204 to the Steam Vehicle Corporation of America." (*Encyclopedia of American, 1805-1942*, First Edition, 1298.) Within a few years, production of Stanley cars ceased altogether.

[445] Cohn, 41-42.

[446] Http://prsteam.inventdata.com.au/steamcar3.html September 21, 2000. "The Failure of the Steam Car."

[447] In late 1913, with the price of gasoline on the rise, the Stanley cars were switched over to white kerosene except for the all-important firing of the boiler. (Katz, 26). Other steamers by then generally used white kerosene, or a close mixture thereof, much like the Stanley.

[448] Yost, *Great Old Cars*, 97.

[449] ibid., 101.

[450] Purdy, *Kings*, 206.

[451] Powell, 202.

[452] James M. Laux, "Steam Cars," *Encyclopedia, 1896-1920,* 428.

[453] Naul, *Specifications Book, 1930-1969,* 398.

[454] Jamison, 90-91.

[455] Ing. J. Tuma, *The Pictorial Encyclopedia of Transport* (New York: Hamlyn Publishers, 1979), 188.

[456] Jamison, 50.

[457] See Table B. Even in those more innocuous times, manufacturers would not offer for long what the public would not deign buy, or even show interest in.

[458] Flink, *America Adopts,* 230.

[459] Rudi Volti, "Why Internal Combustion?" *Invention and Technology* (Fall, 1990), 44-47.

[460] Automobile Manufacturers Association, *Automobiles of America* (Detroit: Wayne State University Press, 1970), 24-25. On the other hand, scientists quickly—and rather prematurely—started to warn of the dangers of exhausting the supplies of petroleum. (Karolevitz, *Old-Time Autos,* 10-11.)

[461] In fact, "Olds carried the assembling of parts from outside suppliers further than anything attempted before." (John A. Heilig, "Handsome Ransom: R. E. Olds and the Birth of the American Automobile," *Automobile Quarterly* 32, no. 3 (Spring 1993), 12). Interestingly, Olds was also one of the few auto pioneers to dabble in all three types. "By 1887, he [Olds] had built. . . [a] three wheeled steam propelled carriage." After that, he tried a number of experimental electrics, before he finally settled on the internal combustion engine because it had, "the most promise." (Dennis Casteele, *The Cars of Oldsmobile* (Sarasota, Florida: Crestline Publishing, 1981), 5.)

[462] Georgano, *Vintage,* 28.

[463] Ruddock, 40.

[464] *Floyd Clymer's Historical Motor Scrapbook,* Vol. 8 (Los Angeles: Clymer Publications, 1955), 19.)

[465] Http://www.hfmgv.org/histories/pic/97.html. Certainly, a small gas tank, if this were the *only* flaw, could be easily rectified. Alas, as we have seen, there were many more.

[466] Flink, *The Automobile,* 7. Surely that one factor *alone* must have eventually doomed the steam car. Especially once the gasoline car got more "user friendly."

[467] "Automobile Notes and Suggestions," *Scientific American* 100 (January 16, 1909), 60.

[468] Shacket, 17.

[469] Georgano, *American Auto 1893-1993,* 73.

[470] Carl Crow, *The City of Flint Grows Up* (New York: Harper Brothers, 1945), 50.

[471] Flink, *America Adopts,* 307.

[472] Janeway, 91.

[473] It might be fitting to record the fate of the 1975 Elcar 2000, one of a plethora of latter-day electrics. The makers tried to endow the car with the latest in technology, but, "for the most part, the technology was little advanced over the Baker. . . of years long ago." (William S. Locke, *Elcar and Pratt Automobiles: The Complete Story* (Jefferson, North Carolina: McFarland & Co., Inc., 2000), 189.)

[474] Jan P. Norbye, *The 100 Greatest American Cars,* (Blue Ridge Summit, Pa: Tab Books, 1981), 192.

BIBLIOGRAPHY.

BOOKS AND PUBLISHED WORKS.

A Biographical Record of Clark County, Ohio. New York: The S.J. Clarke Publishing Company, 1902.

Automobile Manufacturers Association. *Automobiles of America.* Detroit: Wayne State University Press, 1970.

Angelucci, Enzo and Alberto Belluci. *The Automobile from Steam to Gasoline.* New York: McGraw Hill Book Company, 1974.

Bailey, F. Scott and the Editors of Automobile Quarterly. *The American Car Since 1775.* Princeton, New Jersey: Princeton Publications, 1971.

Baldwin, Nick and G. N. Georgano, Michael Sedgwick, and Brian Laban. *The World Guide to Automobile Manufacturers.* New York: Facts on File, 1987.

Bellamy, James F. *Cars Made in Upstate New York.* Red Creek, New York: Squire Hill Publishing Company, 1989.

Bentley, John. *Great American Automobiles: A Dramatic Account of their Achievements in Competition.* New York: Bonanza Books, 1957.

Bentley, John. *Oldtime Steam Cars.* Greenwich, Conn.: Arco Publishing Company, 1953.

Bentley, John. *The Oldtime Automobile.* Greenwich, Conn.: Fawcett Publications, Inc., 1951.

Bergere, Thea. *Automobiles of Yesteryear.* New York: Dodd, Mead & Company, 1962.

Bird, Anthony. *Antique Automobiles*. New York: E. P. Dutton & Co., 1967.

Bixler, Lorin. *Cornelius Aultman, C. Aultman & Co., and the Aultman Company*. Enola, Pennsylvania: Stemgas Company, 1967.

Bolan, Nelson. *Running Board Cars*. Lawrenceville, Virginia: Brunswick Publishing Company, 1987.

Borgeson, Griffith. *"'V, as in Victor, V, as in V8,'" Footnotes to a History of the Racing Darracqs*. Found at NAHC, no date.

Borgeson, Griffith. *The Golden Age of the American Racing Car*. New York: Bonanza Books, 1966.

Cameron, William T. *The Cameron Story*. Tuczon, Arizona: International Society for Vehicle Preservation, 1990.

Cannon, William A. and Fred K. Fox. *Studebaker: The Complete Story*. Blue Ridge Summit, Pennsylvania: Tab Books, 1981.

Casari, Robert B. and Luvada Kuhn, Patricia F. Medert, William H. Nolan., eds. *Chillicothe, Ohio: 1796-1996 Ohio's First Capital*. Jackson, Ohio: Jackson Publishing Company, 1995.

Casteele, Dennis. *The Cars of Oldsmobile*. Sarasota, Florida: Crestline Publishing Company, 1981.

Clark, Donald, ed. *Anatomy of the Automobile: How It's Built— What Makes it Run*. New York: Galahad Books, 1976.

Clymer, Floyd. *Those Wonderful Old Automobiles*. New York: Bonanza Books, 1953.

Clymer, Floyd. *Treasury of Early American Automobiles, 1877-1925*. New York: Bonanza Books, 1950.

Cleveland, Reginald M. and S. T. Williamson. *The Road is Yours-The Story of the Automobile & the Men Behind it.* New York: Greystone Press, 1951.

Cohn, David. *Combustion on Wheels.* Boston: Houghton Mifflin Company, 1944.

Collier, Peter and David Horowitz. *The Fords: An American Epic.* New York: Summit Publishing, 1987.

Conde, John. *Cars with Personalities.* Keego Harbor, Michigan: Arnold Porter Publishing, 1982.

Crow, Carl. *The City of Flint Grows Up.* New York: Harper Brothers, 1945.

Day, John Robert. *The Bosch Book of the Motor Car: Its Evolution and Engineering Development.* New York: St. Martin's Press, 1976.

Derr, Thomas. *The Modern Steam Car and its Historical Background.* Los Angeles: Clymer Publications, 1944.

Donovan, Frank. *Wheels for a Nation.* New York: Thomas Y. Crowell Co., 1965.

Eighth Annual Old-Fashioned Day: Yoctangee Park September 3, 1978. Chillicothe, Ohio: Craftsman Printing, Inc., 1978.

Flammang, James M. and the Editors of Consumer Guide. *100 Years of the American Auto: Millennium Edition.* Lincolnwood, Illinois: Publications International, 2000.

Flink, James J. *America Adopts the Automobile, 1895-1910.* Cambridge, Massachusetts: MIT Press, 1970.

Flink, James J. *The Automobile Age.* Cambridge, Massachusetts: MIT Press, 1988.

Floyd Clymer's Album of Historical Steam Traction Engines and Threshing Equipment No. 1. New York: Bonanza Books, 1959.

Floyd Clymer's Historical Motor Scrapbooks, Vol. 1-8. Los Angeles: Clymer Publications, 1944-1955.

Floyd Clymer's Historical Steam Car Scrapbook. Los Angeles: Clymer Publications, 1945.

Fraser, Richard and Nancy. *A History of Mainer Built Automobiles*, 1834 - 1934. Privately Published, 1991

Freeman, Larry. *The Merry Old Mobiles.* Watkins Glen, New York: Century House, 1949.

Georgano, Nick. (G. N.). *The American Automobile: A Centenary 1893-1993.* New York: Smithmark Publishers, 1992.

Georgano, G. N. (Nick), ed. *The New Encyclopedia of Motorcars 1885 to the Present.* Third Edition. New York: E. P. Dutton, 1982.

Georgano, G. N. (Nick). *Vintage Cars: 1886-1930.* Twickenham, England: Tiger Books, 1977.

Glasscock, Carl Burgess. *Motor History of America.* Los Angeles: Clymer Publications, 1946.

Goddard, Stephen B. *Colonel Albert Pope and His American Dream Machines: The Life and Times of a Bicycle Tycoon Turned Automotive Pioneer.* Jefferson, North Carolina: McFarland & Company, 2000.

Grabb, John R. *Little Known Tales of Old Chillicothe and Ross County, Ohio.* Chillicothe, Ohio: John Webb Printer, 2001.

Greenleaf, William. *Monopoly on Wheels: Henry Ford and the Selden Automobile Patent.* Detroit: Wayne State University Press, 1961.

Heald, E. T. *The Stark County Story.* Canton, Ohio: Stark County Historical Society, 1952.

Heasley, Jerry. *The Production Figure Book For U. S. Cars.* Osceola, Wisconsin: Motorbooks International, 1977.

Helck, Peter. *Great Auto Races.* New York: Henry N. Abrams, 1975.

Helck, Peter. *The Checkered Flag.* New York: Castle Books, 1961.

Henry, Maurice D. *Cadillac: Standard of the World.* Third Edition. Princeton, New Jersey: Princeton Publishing Co., 1979.

Hornung, Clarence P. with James J. Bradley. *100 Great Antique Automobiles.* New York: Dover Publications, 1991.

Jamison, Andrew. *The Steam Powered Automobile.* Bloomington, Indiana: Indiana University Press, 1970.

Janeway, Elizabeth. *The Early Days of Automobiles in America.* New York: Random House, 1956.

Karolevitz, Robert F. *Old-Time Autos in the Ads.* Yankton, South Dakota: The Homestead Publishers, 1973.

Karolevitz, Robert F. *This Was Pioneer Motoring.* Seattle, Washington: Superior Publishing, 1968.

Katzell, Raymond A., ed. *The Splendid Stutz.* Indianapolis, Indiana: The Stutz Club, 1996.

Kimes, Beverly Rae and The Editors of Automobile Quarterly. *Packard: A History of the Motorcar and the Company.* Princeton, New Jersey: Princeton Publishing Company, 1978.

Kimes, Beverly Rae, with Henry Austin Clark, Jr. *Standard Catalog of American Cars 1805-1942.* First Edition. Iola, Wisconsin: Krause Publications, 1985.

Kimes, Beverly Rae. *Standard Catalog of American Cars 1805-1942.* Third Edition. Iola, Wisconsin: Krause Publications, 1996.

Kirsch, David A. *The Electric Vehicle and the Burden of History.* New Brunswick, New Jersey: Rutgers University Press, 2000.

Lewis, Eugene, *Motor Memories.* Detroit: Alred Publishers, 1947.

Locke, William S. *Elcar and Pratt Automobiles: The Complete Story.* Jefferson, North Carolina: McFarland & Company, 2000.

Lutzenberger Picture File, 0300-0449. Main Library. Dayton, Ohio.

MacDonald Prospectus. No date. (Literature for the MacDonald Steamer).

McGaughey, William. *American Automobile Album.* New York: E. P. Dutton, 1954.

Mandel, Leon. *American Cars: From Harrah's Automobile Collection.* New York: Stewart, Tabori & Chang, 1982.

Marvin, Keith, with Arthur Lee Homan. *The Cars of 1923.* Troy, New York: The Automobilists of the Upper Hudson Valley, Inc., 1957.

Marvin, Keith, with Alvin J. Arnheim and Henry H. Blommel. *What was the McFarlan?* New York: Privately Published, 1967.

Matteucci, Marco. *History of the Motor Car.* New York: Crown Publishers, 1970.

Maxim, Hiram Percy. *Horseless Carriage Days.* Boston, Massachusetts: Little Brown & Co., 1937.

Mays, George S., ed. *Encyclopedia of American Business History and Biography: The Automobile Industry, 1896-1920.* New York: Facts On File, 1990.

Medert, Patricia. *Stories from Chillicothe's Past.* Chillicothe, Ohio: Privately Published, 1998.

Melton, James, with Ken Purdy. *Bright Wheels Rolling.* Philadelphia: McRae Smith Company, 1954.

Merksamer, Gregg D. *A History of the New York International Auto Show: 1900-2000.* Atlanta, Georgia: Lionheart Books, Ltd., 2000.

Moloney, James P. *Encyclopedia of American Cars 1930-1942.* Sarasota, Florida: Crestline Publishing Company, 1977.

Musselmann, Morris McNeal. *Get a Horse!—The Story of the Automobile in America.* New York: Lippincott Company, 1950.

Naul, G. Marshall. *The Specification Book for U. S. Cars 1920-1929.* Osceola, Wisconsin: Motorbooks International, 1978.

Naul, G. Marshall. *The Specification Book for U. S. Cars 1930-1969.* Osceola, Wisconsin: Motorbooks International, 1980.

Nicholson, T. R. *Passenger Cars 1863-1904.* New York: MacMillian Company, 1970.

Nicholson, T. R. *Racing Cars and Record Breakers: 1898-1921.* New York: MacMillian Company, 1971.

1903 New York Automobile Show Booklet. New York: Association of Licensed Automobile Manufacturers.

Norbye, Jan P. *The 100 Greatest American Cars.* Blue Ridge Summit, Pennsylvania: Tab Books, 1981.

Oppel, Frank, ed. *Motoring in America: The Early Years.* Secaucus, New Jersey: Castle Books, 1989.

Powell, Sinclair. *The Franklin Automobile Company.* Warrendale, Pennsylvania: Society of Automotive Engineers, 1999.

Purdy, Ken. *The Kings of the Road.* Boston: Little, Brown & Co., 1952.

Purdy, Ken. *Motorcars of the Golden Past.* New York: Galahad Books, 1966.

Purdy, Ken. *The Wonderful World of the Automobile.* New York: Thomas Y. Crowell Co., 1960.

Rae, John Bell. *The American Automobile: A Brief History.* Chicago: University of Chicago Press, 1965.

Riley, Robert Q. *Alternative Cars in the 21st Century: A New Transportation Paradigm.* Warrendale, Pennsylvania: Society of Automotive Engineers, 1994.

Roberts, Peter. *A Picture History of the Automobile.* London: Triune Books, 1973.

Roberts, Peter. *Collector's History of the Automobile.* New York: Bonanza Books, 1978.

Rollins, Ron, ed. *For the Love of Dayton: Life in the Miami Valley 1796-2001.* Dayton, Ohio: Dayton Daily News, 2002.

Schallenburg, Richard. *Bottled Energy: Electrical Engineering and the Evolution of Chemical Energy Storage.* Philadelphia: American Philosophical Society, 1982.

Scharchburg, Richard P. *Carriages Without Horses: J. Frank Duryea and the Birth of the American Automobile Industry.* Warrendale, Pennsylvania: Society of Automotive Engineers, 1993.

Schroeder, Joseph J., Jr. *The Wonderful World of Automobiles 1895-1930.* Chicago: Follet Publishing Co., 1971.

Sedgewick, Michael. *Early Cars.* New York: Octopus Books, 1962.

Shacket, Sheldon R. *The Complete Book of Electrical Vehicles.* Chicago: Domus Books, 1981.

Stein, Ralph. *The American Automobile.* New York: Random House, 1975.

Stein, Ralph. *The Treasury of the Automobile.* New York: Golden Press, 1961.

Stein, Ralph. *The World of the Automobile.* New York: Ridge Press, 1973.

Stern, Phillip Van Doren. *A Pictorial History of the Automobile.* New York: Viking Press, 1953.

The Editors of Automobile Quarterly. *GM: The First 75 Years of Transportation Progress.* Princeton, New Jersey: Princeton Publications, 1983.

Tuma, Ing. J. *The Pictorial Encyclopedia of Transport.* New York: Hamlyn Publishers, 1979.

Wager, Richard. *Golden Wheels: The Story of Automobiles Made in Cleveland and Northeastern Ohio, 1892-1932.* First Edition. Cleveland, Ohio: Western Reserve Historical Society, 1975.

Wagner, Fred J. *Saga of the Roaring Road: A Story of Early Auto Racing in America.* Los Angeles: Clymer Publications, 1949.

Wick, Douglas A. *Automobile History Day-By-Day.* Bismarck, North Dakota: Hedemarken Collectibles, 1997.

Wise, David Burgess. *Classic American Automobiles.* New York: Galahad Books, 1980.

Yost, Stanley K. *The Great Old Cars—Where Are They Now?* Mendota, Illinois: Wayside Press, 1960.

Yost, Stanley K. *They Don't Build Cars Like They Used To.* Mendota, Illinois: Wayside Press, 1963.

About the Author

A lifelong interest in the field of automobiles has always been of special significance for Herbert J. Redman. It remains a curious blend of anticipation and dread to this writer to ponder over what shape the car may take in the near future. This work, Mr. Redman's first book, dwells at great length on what that shape was more than a century ago. And, indirectly, points to what it may it yet become. A member of the Society of Automotive Historians, Mr. Redman resides in Huntington, West Virginia with his wife, Vicki, and their two children.